Reed New Zealand Nature Series

Marine Fishes
of New Zealand 1

Reed New Zealand Nature Series

Marine Fishes
of New Zealand 1

Shoreline and Shallow Seas

Larry Paul

Illustrated by Eric Heath

REED

Front cover: *Top*, Pilchard; *Centre*, Conger eel; *Bottom*, Snapper.
Title page: Seahorse

REED PUBLISHING (NZ) LTD
TE KARUHI TĀ TĀPUI O REED (AOTEAROA)
Established in 1907, Reed is New Zealand's largest book publisher, with over 600 titles in print.
www.reed.co.nz

Published by Reed Books, a division of Reed Publishing (NZ) Ltd, 39 Rawene Road, Birkenhead, Auckland 10.
Associated companies, branches and representatives throughout the world.

ISBN-13: 978 0 7900 0465 5
ISBN-10: 0 7900 0465 8

First published 1997
Reprinted 2001, 2006

Printed in Singapore

Contents

Introduction

New Zealand's northernmost coastline and offshore islands lie in subtropical waters, while its southern offshore islands extend into the subantarctic region. A long coastline has a diversity of habitats and so our our waters contain a large variety of marine fishes. By the mid 1990s about 1100 species were recorded, with new discoveries still occurring.

Marine Fishes of New Zealand 1 and *2* describe most of the commonly seen, caught, or eaten fishes, or (in some cases) at least one representative of a group of very similar species. They also include, for interest, a number of the rarer and unusual fishes that may be seen by fishers or naturalists, or mentioned in other writings on our marine fauna.

This book covers the fishes found predominantly along the shoreline and in shallow seas while *Marine Fishes of New Zealand 2* covers fishes of the deeper coastal and ocean waters. However, the mobility of many fishes means that it is not possible to make a clearcut distinction between 'shallow' and 'deep', or 'inshore' and 'offshore' species.

Deciding whether a fish should be described as 'common', 'uncommon', or 'rare' is also difficult. A fish may be common in deep water but only seen by the commercial fishers who work there, and subsequently in fish shops. Or it may be abundant but small and seldom caught or seen. There are also geographical variations in the abundance of many species. In these books the best estimate of actual – rather than apparent – abundance, is given.

The arrangement of fishes within the texts is by classification (based on anatomical similarities) into families, genera, and species.

These books are not for specialists, but for those who like to fish at weekends and in the evenings; for children who wish to identify the main kinds of fish they find in rock pools; for those who fish in coastal waters; for divers who observe fishes in their natural environment; and for those who appreciate knowing something of the natural world around them.

Fish as successful animals

All fishes plainly show their mastery of the sea. They have evolved in different ways to fill the great variety of marine habitats that exist, and to minimise direct competition between species. They are complex, successful, marine animals. They can remain almost stationary, dart suddenly at their food or swim rapidly away from danger. They can camouflage themselves by quick and subtle changes of colour. Eyes, 'ears', and other sense organs give news of happenings in the surrounding water, and complex social behaviour – often involving schooling – has evolved to meet the emergencies of everyday life, as well as the specialised behaviour during breeding. Fish are very alert to movement, either through their vision, or detection of sound vibrations. Some are able to produce sound themselves. All species occupy their own niche in the interlocking web of marine life. They prey upon each other, as well as on other forms of marine life. In turn, they are a food source for marine mammals and seabirds.

What are 'fish'?

All fishes have heads, bodies, gills, tails, internal organs and skeletons, and most have fins and scales. The shape and arrangement of these parts is important in identifying and classifying the main groups of fishes. There are, broadly speaking, three groups.

The **agnathans** are fish without mouths, the lampreys and hagfishes. New Zealand has a lamprey (mainly freshwater) and one common and two rare hagfishes.

The **cartilaginous** fishes are the sharks, rays, skates, and chimaeras (ghost sharks and elephantfishes). There are more than 600 kinds, about 90 in New Zealand. They have a cartilaginous skeleton, up to seven gill openings, fins that do not fold, no gas bladder, and if scales are present they are small and do not overlap.

The **bony fishes**, or teleosts, are the largest group. There are about 20,000 kinds, more than 1000 in New Zealand. Their features include a hard internal skeleton, a single pair of gill openings, fins which can be raised and lowered, and usually a gas bladder.

Fish biology

Colour

The illustrations in this book have been prepared from colour photographs, field notes and, where possible, from specimens of the fish themselves. They may vary from the fish you see, because the typical colouring of a fish species is often difficult to define. Colour patterns are strongly influenced by habitat.

Juvenile fishes are often more intricately patterned than adults because they must blend with their surroundings. If the adults remain in areas of rock and weed they usually retain a pattern, although it may change to match a new background when the fish moves. When the adults move to more open water they become more uniform in colour.

Although a fish may appear brightly coloured to human eyes, it may be well-camouflaged to the eyes of its underwater predators, matching the bright hues of the encrusted rocks around it. (Bright colours may alternatively be a warning that the fish is poisonous or dangerous.) Iridescence reflects the colour of the surrounding water.

Pelagic fishes tend to be greenish above in coastal seas, bluish in blue ocean water. They have paler sides and undersurfaces, sometimes silver, so they do not appear as a dark shape when seen from below.

Deepwater fish are often dark-coloured, frequently red, brown, or black. These colours are quickly absorbed from the light spectrum as it passes down through water, making these fishes less visible.

Colour is part of a fish's living system, and during the breeding season it may intensify. Conversely, it often fades or alters, and patterns disappear, after death.

Skin and scales

The skin is covered with a layer of mucous which acts as protection against infection and reduces friction when swimming.

Fishes have an external 'skeleton' of scales, which differ greatly in size and shape. They provide protection, and their overlapping pattern allows them to slide over each other, providing body flexibility for swimming. Fishes without scales tend to have a firm, often leathery, skin.

Shark scales (denticles) have spines or ridges which give the skin a rough texture. They are believed to be modified teeth, and strengthen the skin and improve water flow across it.

Fins

Fishes have paired and single fins. The unpaired fins around the body's midline are the dorsal, anal and caudal (tail) fins. The dorsal fin may be divided into two or three fins; spines may support its front portion, or it may contain only soft rays.

Two sets of paired fins (pectoral and pelvic) are usually present. The pectoral fins are on the side of the body just behind the gill opening. In most fish they are used to aid manoeuvering.

The paired pelvic fins are on the underside of the abdomen. They may be back near the vent, forward beneath the pectoral fins, or (occasionally) reduced to a few jaw filaments.

Gas bladder

This is a gas-filled structure in the gut cavity of many bony fish, below the backbone. Fish which regularly move between different depths can quickly adjust the amount of gas in the bladder and compensate for pressure changes. If fish without this ability are forced into shallower water the bladder expands and floats them to the surface.

The lateral line

This is a fine tube beneath the skin which opens to the exterior through holes in the scales. In addition to the main line along the side of the body, the system branches over the head and jaws as rows of pores. This sensory system enables the fish to feel water-pressure changes, and thus detect the 'noise' of nearby prey or predators.

Reproduction

Bony fishes: Coloration sometimes differs, but usually the sex can only be determined from the internal reproductive organs, paired elongated structures just below the backbone. Ovaries are pinkish, granular and rounded. As spawning approaches they enlarge and eggs become visible. The male testes are white or yellowish, smooth and angular.

Most bony fishes produce from thousands to millions of small eggs, usually about 1 mm in diameter. Some build stony nests in which the eggs are deposited in clumps or sheets, some simply scatter their eggs across the seafloor. The majority produce eggs which float towards the surface. The eggs are filled with yolk to nourish the embryo; most hatch within a few days. Tiny transparent larvae remain in the plankton community for days, weeks or months and grow only slowly into miniatures of their parents.

Cartilaginous fishes: The ovary is a cluster of yellow eggs in the front of the body cavity; the testes are longer organs more like those of bony fishes. The eggs enlarge, pass into the oviduct and are fertilised, then move into the uterus where they develop, either into egg cases to be laid or into fully formed young to be born. These fishes produce relatively few young, and their reproductive cycle is often longer than a year.

Distribution patterns

Geographical groupings

Distribution patterns reflect fishes' origins and their preference for either warm or cool water. Warmwater species occur around the North Island, with a subtropical group present at least seasonally from North Cape to East Cape. Coolwater species occur around the South Island, with a few subantarctic species in the far south. Off central New Zealand coolwater species move north and inshore in winter, and warmwater species become more abundant in summer.

It is an over-simplification to categorise northern, southern, and central faunas. Many fishes occur wherever the habitat is suitable; for example, southern species may occur in the north but in deeper water where the water is cool. Some fish wander long distances, apparently tolerant of changing conditions, and others migrate seasonally. The fish community changes with distance from shore and depth, where conditions become more uniform; deepwater species tend to be widely distributed.

Habitats

Another way of categorising marine fishes is by habitat: mangrove harbours and tidal mudflats, sandy bays and beaches, rocky coastlines, inshore reefs, offshore reefs, the broad continental shelf, the steeper outer shelf and continental slope, the latter with canyons, and the clear oceanic water beyond. Two groups of fish are present in each of the habitats: demersal or bottom-dwelling fishes, and pelagic fishes which live mostly in midwater or at the surface.

This book describes the fishes of the shoreline and shallow seas: estuaries, harbours and mudflats, sandy bays and beaches, rocky coastlines and inshore reefs.

We hope that readers find this book useful and informative, and that it stimulates interest in the diversity of marine fishes around New Zealand.

Glossary

Algae: Fresh or saltwater plants: small drifting cells (phytoplankton), fine plants forming a seafloor layer, and large seaweeds.

Anal fin: A fin in the midline behind the anus or vent, and in front of the tail fin.

Barbel: A fleshy tentacle usually below the chin.

Cartilage: Soft bone or gristle; typical of sharks and their relatives.

Caudal: The tail region, as in caudal peduncle (tail base), caudal fin (tail).

Compressed: Flattened from side to side.

Cosmopolitan: Widely distributed through the world.

Crustaceans: Crabs, shrimps, sand-hoppers, sealice, etc.

Dorsal: Upper, as in dorsal surface, dorsal fin.

Endemic: Native to a region, and occurring only there.

Fauna: The group of animals characteristic of a region.

Finlets: Small soft midline fins behind the dorsal and anal fins.

Fin rays: Skeletal rods supporting the fins: they may be of spines or soft rays.

Gill plate: Hard covering over the gill cavity.

Habitat: The particular environment where an organism or group of organisms live; e.g. midwater, underwater caves, the soft seafloor.

Indo-Pacific: The geographical region from the southeast Indian Ocean to the southwest Pacific Ocean.

Lateral line: The row of pores along the side of the fish's body and often extending less clearly over the head; a sensory system.

Operculum: The gill plate or cover.

Oviparous: A reproductive method where eggs develop and hatch outside the female; the eggs, with or without a protective case, are laid on the seafloor or released into midwater.

Ovoviviparous: A reproductive method where eggs develop within the female, nourished by a yolk sac but not directly attached via a placenta.

Paired fins: The pectoral and pelvic fins.

Pectoral fins: Paired fins behind the gills. The extended 'wings' of skates and rays.

Pelagic: Living in the open sea, commonly in the upper layer of the water.

Pelagic tunicates: Small, barrel-shaped, usually transparent planktonic animals; they occur either singly or in aggregations, often chains ('chain jelly'). Salps are a common example.

Pelvic fins: Paired fins on the belly, sometimes called ventral fins.

Plankton: Small single-celled plants (phytoplankton) or small juvenile or adult animals (zooplankton) which drift in midwater.

Scute: An enlarged, strong scale, often ridged; e.g. on the lateral line and tail base of trevally and jack mackerels.

Spiracle: Opening behind the eye in sharks and rays, used in respiration.

Subtropical: The warm region between the tropics and the cooler temperate region.

Temperate: The region between the subtropical and subantarctic (or subarctic) regions.

Trawl: A cone- or funnel-shaped net towed across the seafloor or (less commonly) in midwater by a commercial fishing vessel.

Vent: External opening from the intestine; the anus.

Ventral fins: See Pelvic fins.

Viviparous: A reproductive method where the young are directly nourished by the female via a placenta and born fully developed.

1 Hagfish/Tuere/Napia

Eptatretus cirrhatus

Family MYXINIDAE
Hagfishes

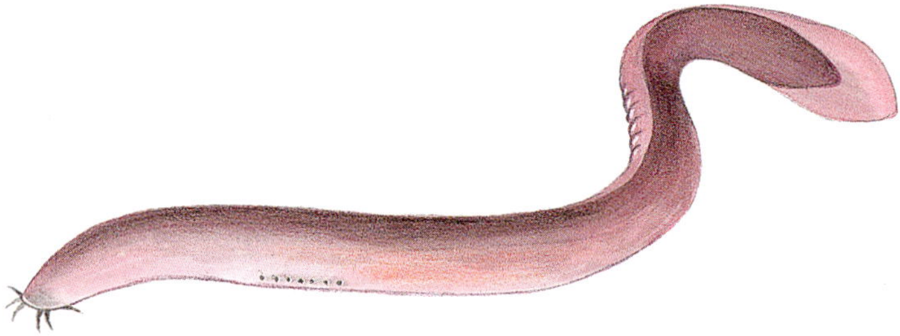

2 Broadsnouted sevengill shark/Tuatini

Notorynchus cepedianus

Family HEXANCHIDAE
Cow sharks

Relationships • Three members of this family occur here; the others, the sixgill and the sharpsnouted sevengill, occur in deeper water.

Other Names • Ground shark (misleading).

Relationships • Three hagfishes occur here; the other two are rare and in deep water.

Other Names • Blind eel.

Distribution • Present in New Zealand waters from the shore to depths of at least 600 m. Less common around northern coasts.

Recognition • The eel-like body form, sucking mouth and circles of teeth, and the ease with which it ties itself in knots and exudes a large volume of slime are unmistakable features. *Colour:* pinkish-brown.

Size • Average length 40–50 cm; maximum 60 cm.

Food • Normally feeds by sucking onto fishes and rasping their body tissues, leaving skin and bones.

Reproduction • Unknown, but like other hagfish probably lays largish eggs.

Capture • *Recreational fishers:* on lines, but seldom if ever deliberately sought. *Commercially:* also accidentally on setlines; very occasionally in trawls. Specimens are taken for dissection in biology classes. The skin can be processed into leather.

Distribution • Widespread in northern and central New Zealand. Frequents coastal waters, sometimes entering shallow bays and harbours. Occurs though the Indo-Pacific at least, with the same or similar species in many other seas.

Recognition • Medium-sized. Head very broad, with a rounded blunt snout, body tapering to a long tail. The single dorsal fin is set far back. Seven pairs of gill slits instead of the usual five. The wide mouth has large oblique teeth; the upper ones have one main point, the lower are broader and comb-like. *Colour:* Pale brown above with many small black and white spots, creamy white below. (The sharpsnouted sevengill is slender and pale grey.)

Size • Average length 150–250 cm; maximum about 3 m.

Food • A variety of other fishes, including smaller sharks and rays.

Reproduction • Ovoviviparous: gives birth to many (up to 80) pups, from eggs which have developed inside the female.

Capture • *Recreational fishers:* by line, occasionally setnet. *Commercially:* accidentally by trawl, or on lines. Usually retaliates aggressively, and though not on record as directly attacking people must be handled carefully.

3 Southern dogfish/Koinga

Squalus acanthias

Family SQUALIDAE
Spiny dogfishes

Relationships • A large family with about 30 species here; others described are (all Book 2): the northern spiny dogfish (No. 2), shovelnose dogfish (No. 3), longnose velvet dogfish (No. 4), lucifer dogfish (No. 5), seal shark (No. 6).

Other Names • Spiny or spiky dogfish, spikeback, spotted dogfish.

---

4 Carpet shark/Pekapeka

Cephaloscyllium isabella

Family SCYLIORHINIDAE
Catsharks

Relationships • About eight catsharks occur here; the other seldom-seen species live in very deep water.

Distribution • Bottom-living, largely restricted to cooler southern waters of New Zealand but ranges north to Hawke Bay and Cape Egmont in winter, from close inshore to depths of about 600 m. Usually more abundant in the central to deeper parts of its range but periodically large schools move into quite shallow water. Present in most seas of the world.

Recognition • Small size, sharp spine in front of each dorsal fin, small teeth, absence of an anal fin. *Colour:* grey-brown with a few large white spots on the back. (The northern spiny dogfish is pale grey-brown without spots, and has large green eyes.)

Size • Average length 70–100 cm; maximum about 120 cm. Females average larger than males.

Food • Wide variety of animals, from bottom-living shellfish and crustaceans to midwater fishes and squid. Voracious feeders.

Reproduction • Ovoviviparous: gives birth to up to 20 pups, from eggs which have developed inside the female.

Capture • *Recreational fishers:* unintentionally on lines and in setnets. *Commercially:* by trawl, line, and setnet. Formerly only a nuisance, although the livers were sometimes used for oil, but greater quantities are now landed for sale. Must be handled cautiously because of the fin spines.

Other Names • Cat shark, swell shark (when caught it may inflate its stomach with air or water), ground shark.

Distribution • Widespread around New Zealand in coastal waters but most common in the south where it comes into shallow bays; in the north, more often caught in deeper water, 100–200 m. Prefers a sandy seafloor, often near rough bottom. Also occurs in southern Australia.

Recognition • Broad flat head, two dorsal fins placed above and behind the pelvic fins, anal fin below the second dorsal. Slit-like eyes, often almost hidden by the eye ridges. Small teeth arranged in a rasp-like pattern. *Colour:* rough-textured skin is golden brown above with darker stripes and blotches, yellowish white below.

Size • Average length 80–120 cm; maximum about 200 cm, with large fish more common in the south.

Food • Crustaceans and small fishes.

Reproduction • Oviparous: produces rectangular horny egg-capsules, each corner extended as tendrils to attach the case to seaweed or a seafloor irregularity.

Capture • *Recreational fishers:* occasionally by line. *Commercially:* accidentally by trawl. Almost completely harmless and usually released alive.

5 Spotted gummy shark/Mango/Pioke

Mustelus lenticulatus

Family TRIAKIDAE
Smoothhounds, houndsharks

6 School shark/Tope/Kapeta

Galeorhinus galeus

Family TRIAKIDAE
Smoothhounds, houndsharks

Relationships • Two members of this family occur here; the other species described is the spotted gummy shark (No. 5).

Relationships • Two members of this family occur here; the other species described is the school shark (No. 6).

Other Names • Rig, spotted dogfish, smoothhound, gummy shark. Sold as lemonfish.

Distribution • Widespread in New Zealand waters, but most abundant in the South Island; well known to those netting in harbours and bays. Depth range is from a few metres to about 200 m, but probably more abundant in 25–100 m. Very similar species occur elsewhere in the world.

Recognition • Slender body. Relatively large dorsal fins, lack of fin spines, small flat teeth. *Colour:* greyish brown above with many white spots, white below.

Size • Average length 70–100 cm; maximum 140 cm.

Food • Mostly crabs, shrimps and small shellfish, which are crushed by the flat teeth. Small fishes are eaten less frequently.

Reproduction • Ovoviviparous: gives birth to up to 23 pups, from eggs which have developed inside the female.

Capture • *Recreational fishers:* by line, setnet, and occasionally beach seine. *Commercially:* target-fished by setnet, also caught by trawl. A good boneless food fish.

Other Names • Sand shark (misleading). Previously listed as *G. australis*.

Distribution • Common around New Zealand from shallow coastal waters to depths of about 200 m, and ranges considerably deeper. Less common offshore, but migrations between New Zealand and Australia occur. Once thought to be only Australasian in distribution; now recognised as the southern form of a species present in many seas of the world.

Recognition • Slender body, flat snout, small sharp teeth, expanded upper tip to the tail margin. *Colour:* greyish brown above, white below.

Size • Average length 120–150 cm; maximum 200 cm.

Food • Wide variety of bottom-dwelling invertebrates (shellfish, crabs, shrimps, worms, etc.); small to medium-sized fishes.

Reproduction • Ovoviviparous: gives birth to 10–35 pups, from eggs which have developed inside the female.

Capture • *Recreational fishers:* usually by line, occasionally in setnets. *Commercially:* target-fished by longline and setnet, accidentally by trawl. The high-quality liver oil was used for vitamin A during the 1940s, but now only the flesh is sold. A good boneless food fish.

7 Short-tailed stingray

Dasyatis brevicaudatus

Family DASYATIDIDAE
Stingrays

Relationships • Three stingrays occur here; the other common black ray is the long-tailed stingray, and there is a seldom-seen pelagic stingray in offshore waters.

---

8 Eagle ray/Whai repo

Myliobatis tenuicaudatus

Family MYLIOBATIDAE
Eagle rays

Relationships • One eagle ray occurs here. Three stingrays and a manta ray are in related families.

Other Names • Yellow ray, sometimes whiptail ray. The Maori

Other Names • Black ray.

Distribution • A predominantly northern species, also found in Cook Strait and sometimes as far south as Otago in summer months. Inhabits shallow coastal areas and bays, found in depths out to 100 m but more common in the shallower parts of its range. Also in Australia and South Africa; quite similar species occur elsewhere.

Recognition • Slightly pointed snout, smooth skin, short tail lacking a dorsal fin, prominent tail spine at about mid-length. *Colour:* dark grey with white spots near the head, white below. (The long-tailed stingray is similar in size but with a longer tail; it can also be distinguished by its brownish grey colour with white spots near the tail base. The front of the pelagic stingray is curved, not pointed, and its body is dark below and above.)

Size • Average length 50–100 cm; maximum about 200 cm.

Food • Mainly shellfish and crabs.

Reproduction • Gives birth to several well-formed young.

Capture • *Recreational fishers:* in beach seines, setnets and on lines fished from beaches. Considered a sport fish when speared. *Commercially:* accidentally in setnets and trawls. (The two coastal stingrays should not be picked up by the head because they can swing the tail and its spine forward to reach this area.)

---

name whai repo is used for this and other stingrays.

Distribution • Widely distributed in New Zealand waters; although more abundant north of Cook Strait it may wander as far south as Otago. Normal habitat is on sandy or sandy mud sediments in harbours and bays and from inshore on sandy coasts to depths of about 100 m. Also around southern Australia; very similar species occur elsewhere.

Recognition • Blunt snout, prominent head, pointed wings (pectoral fins), slender tail, small dorsal fin and spine near the base. *Colour:* olive-brown above, usually with blue markings, white below.

Size • Average length 50–100 cm; maximum about 150 cm.

Food • Shellfish, crabs and fishes are crushed with the flat teeth.

Reproduction • Gives birth to several well-formed young.

Capture • *Recreational fishers:* beach seines, setnets, and lines. *Commercially:* accidentally in setnets and trawls. A nuisance to flounder fishers in shallow northern waters. Because their spine is near the tail base they can be picked up safely by the spiracle holes behind the head (in contrast with other stingrays which can swing their spine forward over the head).

9 Elephantfish/Reperepe

Callorhincus milii

Family CALLORHINCHIDAE
Elephantfishes,
ploughnose chimaeras

10 Longfinned eel/Tuna

Anguilla dieffenbachii

Family ANGUILLIDAE
Freshwater eels

Relationships • New Zealand's other freshwater eel, the shortfin, also has a marine life history stage.

Other Names • There are numerous other Maori names, depending on the eel's size and condition; they also vary regionally.

Relationships • One elephantfish occurs here. It belongs, together with the chimaeras or ghostsharks (Book 2, No. 21) to a group of cartilaginous fishes related to the sharks and rays.

Other Names • Elephant.

Distribution • Well known in southern New Zealand waters, particularly off the Canterbury coast; much less abundant north of Cook Strait. Regularly enters shallow, muddy harbours but is more common in coastal water, and occurs down to about 250 m. There are similar species off southern Australia, South Africa, and South America.

Recognition • Peculiarly shaped fleshy snout with an expanded plough-like tip, large dorsal fin with a strong depressible spine, single small gill slit and smooth silvery skin. *Colour:* silvery grey with darker blotches above, silvery below.

Size • Average length 60–90 cm; maximum 120 cm.

Food • Mainly crabs, shrimps and mud-dwelling shellfish, although small fishes are also eaten.

Reproduction • Lays large horny egg cases, pushed partly into the sandy seafloor, from which well-developed young hatch.

Capture • *Recreational fishers:* setnet, beach seine, occasionally by line. *Commercially:* setnet and trawl. A good boneless food fish.

Distribution • A freshwater eel included here because (in common with the shortfin) it has a marine stage in its life history. Widespread in rivers and streams.

Recognition • A long dorsal fin, extending well forward of the anal fin (to about midway between the latter and the pectoral fin). *Colour:* variable: greyish brown above, paler below. (The shortfin eel is generally smaller, and more common in lowland lakes and streams. Its dorsal fin extends only a short distance forward of the anal fin, and although variable its normal colour is olive green above, paler or white below.)

Size • Average size varies with locality; maximum length 150 cm.

Food • Takes any freshwater animals, as well as fledgling birds, frogs and mice.

Reproduction • Adults migrate to the sea in autumn for spawning, probably north of New Zealand (little is known), while the very small young (elvers) run into estuaries and rivers from the sea during spring.

Capture • *Recreational fishers:* on lines, in trap-nets, by spearing, or simply groped for and caught by hand. *Commercially:* by funnel-shaped fyke-nets. Both species of freshwater eel are highly prized by Maori, and also support valuable commercial fisheries.

11 Yellow moray eel/Puharakeke/ Kaingaaraa

Gymnothorax prasinus

Family MURAENIDAE
Moray eels

12 Conger eel/Koiro/Ngoiro

Conger verreauxi

Family CONGRIDAE
Conger eels

Relationships • Five morays occur here: the yellow, grey, speckled, mottled and mosaic.

Distribution • A warm-water eel found only north of Hawke Bay. Normal habitat is in rocky areas from the shore to depths of about 20 m. The yellow moray is the most common moray eel in New Zealand. It is widely distributed in Australia and probably elsewhere in the Indo-Pacific.

Recognition • Small eyes, teeth and dorsal fin usually hidden by thick skin. *Colour:* variable: yellow or greenish (which fades to brown after death). (The other moray species are well described and almost identifiable by their common names, listed above.)

Size • Average length 80–100 cm; maximum 120 cm.

Food • Crustaceans and small fishes.

Reproduction • Not known; probably open-water spawning to produce planktonic larvae.

Capture • *Recreational fishers:* accidentally on lines. *Commercially:* occasionally on lines or in setnets near reefs, sometimes in lobster pots.

Relationships • Nine congers occur here; most are deepwater or uncommon. The other better-known species include the northern conger (*C. wilsoni*), silver conger (No. 13), and swollenheaded conger (Book 2, No. 24).

Distribution • Inhabits rocky areas to depths of 100 m. Though widespread it is more abundant in central and southern areas. Also known from cooler Australian seas, and there are similar congers elsewhere. (The northern conger occurs around the North Island.)

Recognition • Long dorsal fin, large head and very large mouth. *Colour:* greyish brown above, paler below, becoming darker with increasing size. (The northern conger is very similar but smaller; its dorsal fin starts behind, rather than above, the pectoral fin tip.)

Size • Average length 100–150 cm; maximum 200 cm.

Food • Crustaceans and fishes.

Reproduction • Little-known; an open-water spawner producing planktonic larvae.

Capture • *Recreational fishers:* usually on lines, sometimes speared. *Commercially:* on longlines, sometimes in lobster pots. A good food fish.

13 Silver conger
Gnathophis habenatus

Family CONGRIDAE
Conger eels

14 Pilchard/Mohimohi
Sardinops neopilchardus

Family CLUPEIDAE
Herrings, pilchards, sprats

Relationships • Three herring-type fishes occur here: the pilchard and two sprats (No. 15).

Other Names • Sardine, herring; formerly 'bloater' when commercially canned. The name *S. neopilchardus* is retained here; it has been traditionally used for the Australasian pilchard, but recent studies suggest that pilchard populations in various parts of the world are sufficiently similar to be combined under the name *S. sagax*.

Relationships • Nine congers occur here; most are deepwater and/or uncommon. The other better-known species described are the common conger (No. 12), northern conger (see No. 12), and swollenheaded conger (Book 2, No. 24).

Other Names • Sand eel.

Distribution • A quite common eel of the shallow to central continental shelf. Widely distributed around New Zealand on the sandy to muddy seafloor out to about 100 m. Likely to live in burrows. Also southern Australia.

Recognition • Typical eel shape, with a short pointed snout. Many fishers mistake these eels for small pale juveniles of the true conger. *Colour:* pale olive-grey above, shading to white below, with rather silvery sides; black edges to the fins and tail tip. (The closely related *G. umbrellabia* is very similar but slimmer.)

Size • Average length 30–40 cm.

Food • Probably feeds mainly on small crustaceans.

Reproduction • Not known; probably open-water spawning to produce planktonic larvae.

Capture • *Recreational fishers:* occasionally by line, or in beach seines. *Commercially:* accidentally by trawls.

Distribution • A pelagic coastal fish, widely distributed around New Zealand, but most common in large sheltered embayments such as the Hauraki Gulf and Tasman Bay. Also in southern Australia, with very similar (perhaps the same) species in Japan, South Africa, and off the western coast of North and South America. A common food of seabirds and larger fish whenever they are abundant.

Recognition • The single dorsal fin distinguishes it from small mackerel and mullet, and its colour pattern from other small fishes such as the anchovy and true sprats. *Colour:* blue above, with greenish sides and sometimes a golden iridescence, silvery below. A row of black spots along each side.

Size • Average length 10–20 cm; maximum about 25 cm.

Food • Planktonic crustaceans.

Reproduction • Free-spawning in shallow coastal seas.

Capture • *Recreational fishers:* sometimes pulled ashore in beach seines dragged for flatfish. *Commercially:* can be caught by a variety of netting methods, but only small and sometimes temporary fisheries have developed. The flesh is dark and oily, suitable for smoking or commercial canning.

15 Sprats

Sprattus antipodum and *S. muelleri*

Family CLUPEIDAE
Herrings, pilchards, sprats

Relationships • The pilchard (No. 14) is the only other member of this family here.

---

16 Anchovy/Kokowhawha

Engraulis australis

Family ENGRAULIDIDAE
Anchovies

Other Names • Sardine, herring.

Distribution • Schooling pelagic fishes of coastal waters. Common around the South Island over the shelf, more scattered around the North Island, mainly present in bays. The largest schools east of the South Island are associated with spawning behaviour. They attract feeding concentrations of birds and larger fish. Both are restricted to New Zealand, but similar species of sprat occur elsewhere.

Recognition • Two very similar species, *S. muelleri* slightly deeper-bodied. One dorsal fin. The thin body, and a row of large sharp scales along the belly midline, distinguish the sprats from other small pelagic fishes, which are all plumper. *Colour:* sometimes iridescent, silvery sides and belly. The slender *S. antipodum* tends to be dark blue above; *S. muelleri* is greenish grey with blue patches.

Size • Average length 8–12 cm; maximum 15 cm.

Food • Small planktonic crustaceans.

Reproduction • Free-spawning in shallow coastal seas.

Capture • *Recreational fishers:* occasionally pulled ashore in beach seines dragged for flatfish. *Commercially:* can be caught by a variety of netting methods, but because their schools are mixed with other small pelagic fishes they have not been economic to fish.

Relationships • Only one species occurs here.

Distribution • A pelagic fish of coastal waters. Most common around the North Island and the north-western South Island, generally in sheltered bays but also along open coastlines. Also occurs in Australia. Most commonly encountered in schools, often mixed with sprats and pilchards. All are important prey of pelagic fish and seabirds. Similar anchovy species occur in most seas of the world.

Recognition • One dorsal fin. Distinguished from other similar small pelagic fishes by its large and undershot lower jaw and its *colour:* mainly silver, particularly along the sides, some pale green on the back.

Size • Average length 8–12 cm; maximum 15 cm.

Food • Mainly small planktonic crustaceans.

Reproduction • Free-spawning in shallow coastal seas and large estuaries.

Capture • *Recreational fishers:* entangled and brought ashore in beach seines dragged for flatfish. *Commercially:* can be caught by a variety of netting methods (small-mesh trawl, purse seine, lampara net, etc.), but only small quantities are landed for sale.

17 Sandfish

Gonorynchus gonorynchus

Family GONORYNCHIDAE
Sandfish

18 Ahuru

Auchenoceros punctatus

Family MORIDAE
Morid cods

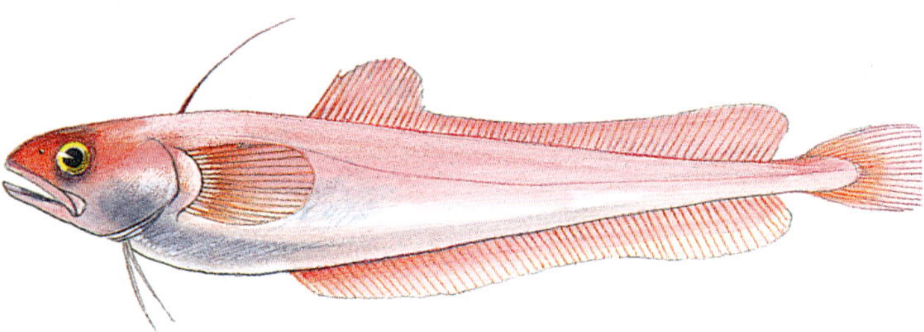

Relationships • A rather primitive fish of uncertain relationships, and the single species in its family.

Other Names • Beaked salmon, sand eel. (Neither is very appropriate; its elongate, bony body and firm scales give it a shark-like or even reptilian appearance.)

Distribution • Widespread around New Zealand, from shallow bays out to the shelf edge, but nowhere common. Occurs from South Africa to Australasia.

Recognition • Sometimes misidentified as a small shark. Body slender, elongate, and rounded, with small, very firm scales. The small mouth without teeth is well back under the pointed snout; dorsal and anal fins are near the rear of the body. *Colour:* greyish brown above, whitish below, with dark patches and white edges on all fins.

Size • Average length 30–40 cm; maximum about 50 cm.

Food • Small bottom-living or burrowing invertebrates, mainly crustaceans.

Reproduction • Unknown, probably free-spawning in coastal waters.

Capture • *Recreational fishers:* occasionally line-caught with a small hook. *Commercially:* accidentally by trawl.

Relationships • About 20 morid cods occur here; others described are the better-known rock cod (No. 19), red cod (No. 20) and ribaldo (Book 2, No. 31).

Distribution • A small cod widespread around New Zealand in shallow waters but localised and apparently sporadic in occurrence and abundance. In some years at least, they are very common in the Firth of Thames and in shallow Otago waters. They appear to occur in schools. Restricted to New Zealand.

Recognition • Essentially codlike in general appearance, with a squarish tail and with the first dorsal fin reduced to a filament. Sometimes mistaken for juvenile red cod, but recognisable by the first dorsal fin and very pale colour. *Colour:* pale pink with a pearly sheen.

Size • Average length 8–10 cm, reaching only about 12 cm.

Food • Planktonic crustaceans.

Reproduction • Unknown; probably free-spawning in shallow coastal waters.

Capture • *Recreational fishers:* occasionally beach seined. *Commercially:* accidentally trapped with other small fishes in trawls.

19 Rock cod
Lotella rhacinus

Family MORIDAE
Morid cods

20 Red cod/Hoka
Pseudophycis bachus

Family MORIDAE
Morid cods

Relationships • About 20 morid cods occur here; most are deepwater and/or uncommon. The other better-

Relationships • About 20 morid cods occur here; most are deepwater and/or uncommon. The other better-known species described are the ahuru (No. 18), red cod (No. 20), and ribaldo (Book 2, No. 31).

Distribution • A codfish which inhabits reefs and rough ground, rather than the open ground of the related red cod and bastard red cods. Widespread and quite common in this habitat around New Zealand, from shallow water down to perhaps 30 m, more common in the south. Also around southern Australia. (In northern New Zealand, there is a similar but uncommon species, the beardie *L. phycis*.)

Recognition • Large head and plump body, tapering to a small rounded tail. Small first dorsal fin, and prominent chin barbel. *Colour:* mid brownish to dark reddish brown, usually with pale margins to the long dorsal and anal fins and tail. No dark blotch by the pectoral fin. (*L. phycis* is smaller [to 20 cm] and smaller-scaled.)

Size • Average length 25–35 cm; maximum at least 40 cm.

Food • Small fishes, crustaceans, and various bottom-living invertebrates.

Reproduction • Unknown; probably free-spawning in shallow coastal waters.

Capture • *Recreational fishers:* by line, setnet, or spear. *Commercially:* accidentally by setnet.

---

known species described are the ahuru (No. 18), rock cod (No. 19), and ribaldo (Book 2, No. 31). Two species of *Pseudophycis* rather similar to the red cod are known as bastard red cods.

Distribution • Present in all but the most northern waters, but most abundant around the South Island. Main habitat is muddy seafloor, from shallow harbours out to depths of at least 400 m. Red cod vary greatly in distribution and local abundance from year to year. Also present around southern Australia.

Recognition • Soft plump body, short first dorsal fin, long second dorsal and anal fins with dark edges, square tail tip. *Colour:* pinkish to reddish brown above, with a black blotch mark at the pectoral fin base, white below. (The bastard red cods are also plumpish and reddish brown but have small rounded tails.)

Size • Average length 40–70 cm; maximum 80 cm.

Food • Seafloor invertebrates detected with the chin barbel, and small fishes.

Reproduction • Free-spawning in coastal waters.

Capture • *Recreational fishers:* by line, beach seines, and setnets. *Commercially:* by trawl and setline. A good food fish.

21 Orange clingfish

Diplocrepis puniceus

Family GOBIESOCIDAE
Clingfishes

22 Garfish/Ihe/Takeke

Hyporhamphus ihi

Family HEMIRHAMPHIDAE
Halfbeaks

Relationships • About a dozen clingfishes occur here.

Other Names • Suckerfish.

Distribution • One of the commoner New Zealand species. Widespread in New Zealand, but absent from warm northern waters, it occupies a very narrow zone immediately above and below low-tide mark along rocky shores. Restricted to New Zealand.

Recognition • A broadly rounded head, short, slender body, and large adhesive disc on the underside; has the habit of clinging to the undersurfaces of stones. *Colour:* may vary through different shades of brown, yellow, green, or pinkish, often with stripes or spots.

Size • Average length about 7–9 cm; maximum 12 cm.

Food • Small crustaceans and shellfish, caught during brief excursions from cover.

Reproduction • Several females contribute to a sheet of eggs laid beneath rocks or in a crevice, which is then guarded by a male.

Capture • Taken carefully by hand from the rock surface.

Relationships • Two garfishes occur here; the other is quite uncommon.

Other Names • Piper, halfbeak.

Distribution • Throughout New Zealand in surface and sub-surface coastal waters. Restricted to New Zealand but with similar species in other seas. Taken as food by many seabirds and by other marine fishes.

Recognition • Slender body, extended lower jaw, rearward-set small dorsal fin. *Colour:* bluish green above, paler below, with silvery sides. (The green or longfinned garfish *Euleptorhamphus viridis* is larger and has a long pointed pectoral fin.)

Size • Average length 20–35 cm; maximum 45 cm.

Food • Mostly small pelagic crustaceans, but in estuaries they will feed on seagrass and fine seaweeds.

Reproduction • Free-spawning in sheltered bays and estuaries, where the eggs become attached to seaweed.

Capture • *Recreational fishers:* in beach seines, on very light lines or dip-netted under a light at night. *Commercially:* in beach seines. A popular food fish.

23 Common roughy/Patohe

Paratrachichthys trailli

Family TRACHICHTHYIDAE
Roughies, slimeheads

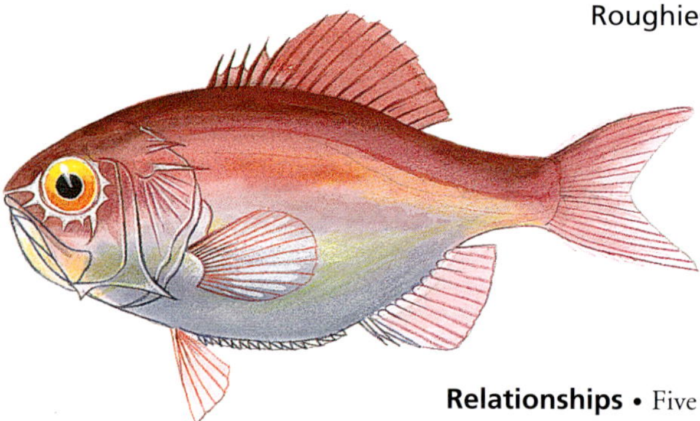

Relationships • Five roughies occur here, including the slender roughy *Optivus elongatus*; others described are (Book 2) the orange roughy (No. 42) and silver roughy (No. 43).

---

24 John dory/Kuparu

Zeus faber

Family ZEIDAE
Dories

Relationships • Six true dories occur here, plus a related small dory-like fish. Other species described are (Book 2) the capro dory (No. 46),

Other Names • Sandpaper fish.

Distribution • Widespread but more common around central and southern New Zealand, to depths of at least 100 m. In shallow water it occurs on reefs, usually sheltering by day and feeding at night; in deep water it occurs over more open seafloor. Restricted to New Zealand, with a very similar species in southern Australia.

Recognition • Large head with bony ridges, large eyes, vent between pelvic fins, with row of scutes between it and the anal fin, rough scales. *Colour:* pinkish bronze, with a silvery sheen on the sides; pink fins. (The slender roughy has large eyes and mouth, and scutes along the midbelly. More slender than the common roughy and distinguishable by having its vent in the normal position near the anal fin. It is pinkish brown with a thin dark stripe on each tail lobe.)

Size • Average length 15–20 cm; maximum 25 cm.

Food • Midwater plankton, particularly crustaceans.

Reproduction • Unknown; presumably free-spawning in coastal waters.

Capture • *Recreational fishers:* occasionally by line and small hook, or speared. *Commercially:* accidentally in setnets, lobster pots, or by trawl.

---

mirror dory (No. 47), and lookdown dory (No. 48).

Distribution • Widespread about the North Island, becoming less common about Cook Strait, and uncommon further south. A range of habitats, but mainly on the open sandy seafloor, from shallow bays to about 200 m. Also occurs in the eastern Atlantic from Britain to South Africa, and in the Indo-Pacific from Australasia to Japan.

Recognition • Almost oval, very narrow body, protrusible mouth, strong spines around much of the body margin. *Colour:* sometimes whitish to pale grey or brown, with a series of curved dark brown horizontal stripes, sometimes almost uniformly olive brown or bronze with a yellow to golden sheen. A prominent midbody black blotch ringed with pale yellow.

Size • Average length 30–45 cm; maximum 60 cm.

Food • Mainly fishes, also large midwater crustaceans, stalked and captured in the suddenly extended jaws.

Reproduction • Free-spawning in coastal waters.

Capture • *Recreational fishers:* by line, setnet or spear. *Commercially:* usually by trawl. A popular food fish.

25 Smooth pipefish

Stigmatopora macropterygia

Family SYNGNATHIDAE
Seahorses, pipefishes

26 Shortnosed pipefish

Lissocampus filum

Family SYNGNATHIDAE
Seahorses, pipefishes

Relationships • Six pipefishes occur here; the other common species described is the shortnosed pipefish (No. 26). The spiny seadragon (*Solegnathus spinosissimus*) is a large and spectacular reddish-orange pipefish of the open ocean.

Distribution • Widespread from Cook Strait southwards among rock and weed in sheltered places below low-tide mark. Usually lives in crevices or in holes in seaweed bases. A similar species occurs in northern waters. Restricted to New Zealand.

Recognition • Slender snout, long dorsal fin and slender, pointed tail without a tail fin. *Colour:* variable: yellowish grey to dark brown.

Size • Average length 25–30 cm; maximum 38 cm.

Food • Small planktonic crustaceans.

Reproduction • Male and female pair, fertilised eggs are passed into a pouch on the male's belly; when larvae hatch they are released into sheltered water.

Capture • Taken by hand in seaweed. Sometimes kept in aquaria.

Relationships • Six pipefishes occur here; the other common species described is the smooth pipefish (No. 25). The spiny seadragon (*Solegnathus spinosissimus*) is a large and spectacular reddish-orange pipefish of the open ocean.

Other Names • Shortsnout pipefish.

Distribution • Throughout New Zealand waters among rock and weed at, and just below, low-tide mark. Restricted to New Zealand.

Recognition • The habitat, short snout, small short dorsal fin and tail fin identify this species. *Colour:* variable, darkish and mottled.

Size • Average length about 10 cm; maximum 12 cm.

Food • Small planktonic crustaceans.

Reproduction • Male and female pair, fertilised eggs are passed into a pouch on the male's belly; when larvae hatch they are released into sheltered water.

Capture • Taken by hand in seaweed. Sometimes kept in aquaria.

27 Seahorse/Manaia

Hippocampus abdominalis

Family SYNGNATHIDAE
Seahorses, pipefishes

---

28 Sea perch/Pohuiakaroa

Helicolenus percoides

Family SCORPAENIDAE
Scorpionfishes

Relationships • Over a dozen scorpionfishes occur here, but only a few are common. Other inshore species described are the red scorpionfish (No. 29) and red rock cod (No. 30); the deepwater sea perch *H. barathri* (Book 2, No. 53) is offshore.

Relationships • One species of seahorse occurs here, and six pipefishes. Pipefishes described are the smooth pipefish (No. 25) and shortnosed pipefish (No. 26).

Distribution • Widespread around New Zealand, although is probably more abundant and greater in size in southern waters. Also occurs in cool Australian seas, and similar species are known throughout the world.

Recognition • Well known for its distinctive 'horse'-like appearance. The Maori name manaia refers to the spiral tail. *Colour:* variable: usually brownish, sometimes more yellow, orange, or red to match different surroundings such as sponges.

Size • Body extended, average length 15–20 cm; maximum 30 cm.

Food • Small weed-dwelling or planktonic crustaceans.

Reproduction • Males, like the pipefish, have the habit of pairing with a female and then carrying developing eggs and very small young in a special abdominal pouch.

Capture • Can be caught by hand from the seaweed in which it lives, or in beach seines which drag seaweed ashore.

Other Names • Jock Stewart, scarpee, Maori chief, fivefinger and scrodie are some of the many alternative names in use, although these are often also used for the related scorpionfishes living in the same reef habitat. ('Sea perch' is often used for fish in other families.)

Distribution • Throughout New Zealand waters, mostly south of the Bay of Plenty, from the rocky shore to depths of 100 m; most abundant in the middle part of this range. Also in cool Australian seas.

Recognition • Head not prominently large and spiny, body shape normal. *Colour:* variable, usually reddish orange and brown, sometimes yellowish; 5 darker vertical bands on the body. (The red scorpionfish and red rock cod have a large head with strong spines, a tapering body, and irregular colour mottling. The deepwater sea perch has larger eyes and is brighter red.)

Size • Average length 25–30 cm; maximum 35 cm.

Food • Moderately large seafloor invertebrates, particularly crustaceans; also smallish fishes.

Reproduction • Ovoviviparous, giving birth to small larvae within a jelly mass; when free these become planktonic.

Capture • *Recreational fishers:* on lines or in setnets over rough ground. *Commercially:* accidentally in setnets.

29 Red scorpionfish
Scorpaena papillosus

Family SCORPAENIDAE
Scorpionfishes

Relationships • Over a dozen scorpionfishes occur here, but only a few are common. Other inshore species described are the sea perch

---

30 Red rock cod/Matuawhaapuku
Scorpaena cardinalis

Family SCORPAENIDAE
Scorpionfishes

(No. 28) and red rock cod (No. 30); the deepwater sea perch (Book 2, No. 53) is offshore.

Other Names • Dwarf scorpionfish. Other names such as cobbler and scarpee tend to confuse this fish with related or similar reddish or orange-brown reef species.

Distribution • Widespread around New Zealand coasts, from the shore down to about 50 m, almost exclusively over rocky ground and reefs. Well-camouflaged, secretive and nocturnal, and not often seen. Also in cool Australian seas.

Recognition • Large, strongly spined head, deep but relatively small tapering body. *Colour:* variable, mainly reddish, mottled with pink, brown, and pale yellow. (The red rock cod is larger, with areas of pale mottling. The sea perch, and unrelated redbanded perch, have less spiny heads and more uniform colour banding.)

Size • Average length 15–20 cm; maximum about 25 cm.

Food • Seafloor invertebrates, particularly crustaceans, occasionally small fishes.

Reproduction • Unknown; possibly ovoviviparous, giving birth to small larvae.

Capture • *Recreational fishers:* by line over rocky ground. The head and fin spines can inflict painful stab wounds.

Relationships • Over a dozen scorpionfishes occur here, but only a few are common. Other inshore species described are the sea perch (No. 28) and red rock cod (No. 29); the deepwater sea perch (Book 2, No. 53) is offshore.

Other Names • Northern scorpionfish, grandfather hapuku.

Distribution • North Cape to East Cape, on reefs or rocky ground, from shallow water down to 100 m. Also in eastern Australia.

Recognition • Large, strongly spined head, deep but relatively small tapering body. *Colour:* variable, mainly reddish to orange brown, with pink and whitish banding and mottling. (The red scorpionfish is smaller, brighter coloured, without areas of pale mottling. The sea perch, and unrelated redbanded perch, have less spiny heads and more uniform colour banding.)

Size • Average length 25–45 cm; maximum 60 cm.

Food • Large crustaceans and smallish fish.

Reproduction • Ovoviviparous, giving birth to small larvae.

Capture • *Recreational fishers:* by line over rocky ground, or speared. *Commercially:* sometimes accidentally in setnets. The head and fin spines can inflict painful stab wounds.

31 Southern pigfish

Congiopodus leucopaecilus

Family CONGIOPODIDAE
Pigfishes, horsefishes

Southern pigfish

Alert pigfish

32 Red gurnard/Kumukumu

Chelidonichthys kumu

Family TRIGLIDAE
Gurnards, searobins

Relationships • Three pigfishes occur here; the others are a larger species in deeper water (*C. coriaceus*), and a slender species in midwater, the alert pigfish (*Alertichthys blacki*).

Distribution • Southern New Zealand, just reaching Cook Strait. Present at depths down to 100 m, but more common in shallower coastal waters, usually but not always near rocky ground or weed beds. Apparently restricted to New Zealand, with a similar species in southern Australia.

Recognition • Large head and pointed snout, tapering to a narrow body, stiffly erect spiny dorsal fin, leathery skin instead of scales. *Colour:* blotched light and dark brown, with short vertical stripes or bands along the flank. (*C. coriaceus* is greyish with a horizontal white stripe. The alert pigfish is large-eyed, smallish, rather stiff-bodied, with a pointed head and snout; it is greyish white with dark grey stripes and bands.)

Size • Average length 20–25 cm; maximum 30 cm.

Food • Shellfish, crustaceans, and marine worms, often extracted from crevices using the slightly extendable snout.

Reproduction • Little-known, but presumably free-spawning in coastal waters.

Capture • *Recreational fishers:* by line and small hook, or setnet. *Commercially:* accidentally by trawl.

Relationships • Four gurnards occur here; others described are (Book 2) the spotted gurnard (No. 54) and scaly gurnard (No. 55).

Other Names • Gurnard.

Distribution • Widespread on open sandy seafloors to depths of 150 m. Also in cool Australian waters and in South Africa, with a very similar species in Japan.

Recognition • Bony head with several small spines, round tapering body, very large colourful pectoral fins, with the first few rays enlarged and used both for 'walking' and sensing buried prey. *Colour:* pale brownish pink with darker pink blotches, white below; becomes more uniformly pink when caught. Pectoral fins dark olive green with a blue margin and spots, a black blotch with white spots near the base.

Size • Average length 30–50 cm; maximum 60 cm.

Food • A variety of seafloor invertebrates, also small fishes.

Reproduction • Free-spawning in coastal waters.

Capture • *Recreational fishers:* line and beach seine. *Commercially:* by line and trawl. Often grunts when taken from the water. A popular food fish.

33 Redbanded perch

Hypoplectrodes huntii

Family SERRANIDAE
Sea basses, groupers

---

34 Olive rockfish/Taumaka

Acanthoclinus fuscus

Family ACANTHOCLINIDAE
Rockfishes, spiny basslets

Relationships • Five rockfishes occur here.

Other Names • Formerly only one rockfish was recognised, usually listed as *A. littoreus*.

Relationships • About 20 small groupers occur here; others described are (Book 2) the spotted black groper (No. 58), pink maomao (No. 59), orange perch (No. 60), and butterfly perch (No. 61).

Other Names • Banded perch. Previously listed as *Ellerkeldia*.

Distribution • Widespread on rough ground in depths from 20–100 m. Restricted to New Zealand. (A similar species of *Hypoplectrodes*, with redder stripes including a horizontal one through the eye, occurs in north-eastern waters, particularly at islands and headlands.)

Recognition • Body rather hump-backed, head pointed. *Colour:* reddish orange above, yellowish to white below, with 7 vertical brown bands. (The unrelated sea perch [No. 28] is spinier, and not as conspicuously banded.)

Size • Average length 25–35 cm; maximum 45 cm.

Food • Seafloor invertebrates and small fishes.

Reproduction • Paired spawning, producing pelagic eggs.

Capture • *Recreational fishers:* by line, sometimes setnet.

Distribution • One of the most distinctive of all New Zealand's small intertidal fishes, occurring throughout the country in rocky places and under scattered stones on mudflats exposed at low water. There are four very similar species of rockfish in the intertidal, and one in coastal waters; this is the most widespread, but others may be more common in some localities. All are restricted to New Zealand.

Recognition • A long body and long, even dorsal fin. Three main lateral lines, the lower subdivided at midbody to form a fourth. *Colour:* greyish to olive green, with a white head stripe. (The other rockfishes have an undivided lower lateral line.)

Size • Average length 20–25 cm; maximum 30 cm.

Food • An aggressive, predatory species of rock pools, the rockfish is the scourge of all animals, including small fishes, living in the vicinity. Prey is stalked slowly, only the pectoral fins being used for movement. At other times swimming is eel-like.

Reproduction • An egg mass is deposited beneath a boulder, the male creating a sealed 'nest' and guarding it until the eggs hatch and larvae leave.

Capture • Can be caught by hand (though very agile and slippery) from under rocks, or on small hooks fished just below low-tide mark.

35 Trevally/Ara-ara

Pseudocaranx dentex

Family CARANGIDAE
Trevallies, kingfishes, jacks

Relationships • About a dozen members of this family occur here; the others described are the yellowtail kingfish (No. 36), jack mackerels (No. 37), and pilotfish (Book 2, No. 64).

---

36 Yellowtail kingfish/Haku

Seriola lalandi

Family CARANGIDAE
Trevallies, kingfishes, jacks

Other Names • Formerly listed as *Caranx lutescens* or *C. georgianus.*

Distribution • Most abundant in northern waters but also found around the north of the South Island, from shallow bays out to 100 m; most often over the open seafloor but sometimes near reefs. Range from the bottom up into midwater, and often seen schooling and feeding at the surface. Widespread in most warm seas of both hemispheres.

Recognition • Oval and rather narrow streamlined body, long curved pectoral fin, stiff forked tail on a narrow tail base bearing a row of bony scutes on each side. Large fish may have a humped head. *Colour:* pale silvery green above, slightly iridescent, silvery sides and belly, yellowish tail. A dark spot on the rear corner of the gill cover. Small juveniles have some vertical darker banding, and slightly larger fish a yellow midbody stripe.

Size • Average length 30–50 cm; maximum 75 cm.

Food • Wide variety of small seafloor and midwater invertebrates, particularly crustaceans.

Reproduction • Free-spawning in coastal waters.

Capture • *Recreational fishers:* by line, setnet, and beach seine in some localities. *Commercially:* by trawl, purse seine, and setnet. A good food fish.

Relationships • About a dozen members of this family occur here; the others described are the trevally (No. 35), jack mackerels (No. 37), and pilotfish (Book 2, No. 64).

Other Names • Northern kingfish.

Distribution • Widely distributed in North Island and northern South Island waters, most common in the north, from shallow bays and harbours out to about the shelf edge. Pelagic, living mainly in midwater and at the surface. Occurs throughout warm temperate waters of the Southern Hemisphere, with close relatives elsewhere.

Recognition • Streamlined torpedo-shaped body, low first dorsal fin, small scales, small tail base scutes. *Colour:* bluish green above, silvery white sides and belly; yellow band from the eye to the tail base, yellowish tail, particularly near the tips.

Size • Average length 60–120 cm; maximum 180 cm.

Food • Pelagic fishes and squid, also small reef fishes.

Reproduction • Free-spawning in open coastal waters.

Capture • *Recreational fishers:* usually taken by trolling with live bait and lures, also speared. *Commercially:* trolling, setnet, purse seine, and trawl. A popular food fish.

37 Jack mackerels/Hauture

Trachurus spp.

Family CARANGIDAE
Trevallies, kingfishes, jacks

Trachurus declivis

Trachurus novaezelandiae

Relationships • About a dozen members of this family occur here; the others described are the trevally (No. 35), yellowtail kingfish (No. 36), and pilotfish (Book 2, No. 64).

Other Names • Horse mackerels. Occasionally (and confusingly) yellowtail for small fish.

Recognition • The streamlined form, greenish colour, long pectoral fin, a lateral line which dips markedly below the second dorsal fin, and scutes (broad and overlapping scales along the lateral line, expanded into a hard ridge at the tail base) are features of all species. They are difficult to separate; key features are the number and size of scutes, and length of an inconspicuous secondary lateral line beside the dorsal fins. Shape and colour are variable but can sometimes be useful. (Another jack mackerel, the koheru [*Decapterus koheru*] is often seen schooling near reefs in northern New Zealand. Distinguished by its plumper body, short pectoral fin, bright bluish green colouring with a yellow stripe becoming more prominent towards the tail, a lateral line which only dips slightly down below the second dorsal fin, and small scutes only along the tail base.)

Food • Midwater invertebrates, particularly crustaceans, and small fishes.

Reproduction • Free-spawning in coastal or ocean waters.

Capture • *Recreational fishers:* usually in setnets, sometimes by line. *Commercially:* different regional fisheries take one or more species together, by purse seine, trawl, and setnet. They are good food fishes, and often commercially canned.

---

1. *T. declivis*

Distribution • Widely distributed, but more common around central and north-western New Zealand. Also in southern Australian seas. Pelagic, mainly over the central shelf.

Recognition • Largish, moderately slender, 71–89 large scutes, secondary lateral line reaches about eighth ray of second dorsal fin. *Colour:* dark greenish blue above, white below.

Size • Average length 35–50 cm; maximum about 55 cm.

2. *T. novaezelandiae*

Distribution • Widely distributed, but more common along New Zealand's north-east coast. Also in southern Australian seas and South-east Asia. Pelagic over the inner to central shelf. Usually the commonest small jack mackerel in harbours and bays.

Recognition • Generally smaller and slightly deeper-bodied than the other two species, 67–81 large scutes, secondary lateral line only reaches first

Trachurus murphyi

38 Kahawai

Arripis trutta

Family ARRIPIDAE

Kahawai, Australian salmon

or second ray of second dorsal fin. *Colour:* green above, smaller fish with faint brownish/yellowish green vertical bands, white below.

Size • Average length 30–40 cm; maximum about 45 cm.

3. *T. murphyi*

Distribution • A South American species (Peruvian, South Pacific, or slender mackerel), first found here in the early 1980s, presumably having extended its distribution westwards across the South Pacific Ocean. Occurs offshore around central and southern New Zealand. Increased in abundance in the early 1990s, becoming the dominant jack mackerel in some areas.

Recognition • Largish, with a slender rear body, 95–106 small scutes. *Colour:* dark bluish green above, white below.

Size • Average length 35–55 cm; maximum 60 cm.

Relationships • Two kahawai species occur here.

Distribution • Widespread around New Zealand, from estuaries out to about mid-shelf in midwater and at the surface. Also occurs in south-eastern Australian waters. (*A. xylabion* occurs in northern New Zealand and at the Kermadecs.)

Recognition • Streamlined, plump but still flat-sided body, smallish eyes. *Colour:* greenish blue above, usually darker in big fish, with various-sized spots in irregular rows, yellow pectoral fins, silvery to white lower sides and belly. Juveniles: pale brownish or golden vertical bars or spots on the side. (*A. xylabion* has a large tail longer than the head [in the common kahawai it is equal to or shorter than the head], and often more prominent eyes.)

Size • Average length 40–50 cm; maximum about 60 cm.

Food • Mainly midwater plankton, particularly crustaceans, and small fishes; also a variety of seafloor invertebrates.

Reproduction • Free-spawning in coastal waters.

Capture • *Recreational fishers:* caught by most methods, particularly lines (trolling, surfcasting, driftlining, etc.), setnet, and beach seine. *Commercially:* majority by purse seining, also setnets, and accidentally by trawl. A good food fish, often commercially canned.

39 Snapper/Tamure

Pagrus auratus

Family SPARIDAE
Seabreams

Relationships • The only seabream here.

Other Names • Formerly listed as *Chrysophrys auratus*. Bream or brim was occasionally used for small fish.

---

40 Red mullet/Ahuruhuru

Upeneichthys lineatus

Family MULLIDAE
Goatfishes

Relationships • Four goatfishes occur here.

Other Names • Goatfish.

Distribution • North Island and warmer parts of the South Island, in a wide variety of habitats from shallow harbours to depths of about 150 m. Also in Australia, and the same or an extremely similar species is the red sea bream of Japan and some coastal Indo-Pacific regions.

Recognition • Large head and strong teeth, but a rather generalised fish in shape. *Colour:* golden pink, many bright blue spots on the upper sides and a blue mark above the eye; white below. Large fish living near reefs become dark red, while those on a shallow muddy seafloor are a paler pink.

Size • Average length 30–50 cm; maximum 100 cm. In general, west coast fish are larger than those on the north-east coast.

Food • Wide variety of seafloor invertebrates, with hard-shelled animals taken by larger fish; also pelagic salps, and smallish fishes.

Reproduction • Free-spawning in coastal waters.

Capture • Probably New Zealand's most sought-after fish, by both commercial and sport fishers. *Recreational fishers:* by line, setnet, beach seine and occasionally (because of its wariness) by spear. *Commercially:* by trawl, line, Danish seine, and setnet. A popular food fish.

Distribution • Found mainly in northern coastal waters, present around the South Island but uncommon south of Cook Strait. Main habitat is on sandy seafloor near reefs. Also south-eastern Australia. (The yellow goatfish occurs at Kermadec Is, the blackspot and bar-tailed goatfishes in northern New Zealand.)

Recognition • Large head, slightly humped behind the eye, two prominent chin barbels, two large and separated dorsal fins. *Colour:* variable, usually bright red with blue stripes and spots, sometimes also with light and dark brown lines and vertical bars. At night they are reddish orange with white markings. Also change colour and pattern when disturbed, and during breeding. (The yellow, blackspot and bar-tailed goatfishes are characterised by the colour pattern their names describe.)

Size • Average length 15–30 cm; maximum 45 cm. Males grow larger than females.

Food • Crustaceans and marine worms, searched out in the sand with the sensitive chin barbels.

Reproduction • Paired spawning in shallow water, producing pelagic eggs.

Capture • *Recreational fishers:* line and small hook, setnet, or speared. *Commercially:* setnet, occasionally by trawl. A good food fish.

41 **Bigeye**

Pempheris adspersus

Family PEMPHERIDAE
Bulleyes, sweepers

42 **Drummer**

Kyphosus sydneyanus

Family KYPHOSIDAE
Drummers, sea chubs

Relationships • Three bulleyes occur here, but only this species is common.

Distribution • Widespread on rocky coastlines, but much more abundant from the Bay of Plenty northwards, from shallow water down to about 70 m. Restricted to New Zealand.

Recognition • Aptly named; the large eyes and mouth dominate the small head, and the body is deep but narrow. *Colour:* coppery brown with faint iridescence, dark tips to dorsal and anal fins and tail, particularly in small fish. Juveniles are pale, almost translucent green.

Size • Average length 5–15 cm; maximum 17 cm.

Food • Adult bigeyes are cave dwellers during the day, and at night emerge to feed mainly on swarms of planktonic animals, sometimes moving into such shallow areas as mangrove estuaries. Juveniles feed during the day.

Reproduction • Little-known, but presumably free-spawning in shallow water producing pelagic eggs.

Capture • Only accidentally in setnets, more often observed and photographed by divers.

Relationships • About eight drummers occur here; the others described are the mado (No. 43), bluefish (No. 44), parore (No. 45), and blue maomao (No. 46).

Other Names • Silver drummer.

Distribution • North Island, most common from North Cape to East Cape. Usually in shallow turbulent water, to 10 m, over rocky ground, but may occur to about 30 m in the clear water around offshore islands. Also in Australia.

Recognition • Stout and almost oval body, small head and mouth. *Colour:* greenish grey above, silvery below, with darker stripes along each scale row.

Size • Average length 30–50 cm; maximum 75 cm.

Food • Almost exclusively seaweed; small weed-dwelling invertebrates are probably eaten accidentally.

Reproduction • Unknown, presumably free-spawning in shallow water producing pelagic eggs.

Capture • *Recreational fishers:* by setnet or speared. *Commercially:* accidentally by setnet.

43 Mado

Atypichthys latus

Family KYPHOSIDAE
Drummers, sea chubs

44 Bluefish/Korokoropounamu

Girella cyanea

Family KYPHOSIDAE
Drummers, sea chubs

Relationships • About eight drummers occur here; the others described are the drummer (No. 42), bluefish (No. 44), parore (No. 45), and blue maomao (No. 46).

Distribution • North Cape to the Bay of Plenty, over rocky ground and reefs, generally where there is cave or crevice shelter. Shallow water down to about 50 m, but generally not common. A similar species occurs in Australia, and both are present at Tasman Sea islands.

Recognition • Small head and almost oval body. *Colour:* white with 5 or 6 dark brown horizontal stripes, the largest of these with a central yellow line; yellow fins.

Size • Average length 20–25 cm; maximum 30 cm.

Food • Small seafloor invertebrates and encrusting organisms. Juveniles also behave as cleaner fish.

Reproduction • Little-known, but presumably free-spawning in shallow water producing pelagic eggs.

Capture • Observed and photographed but seldom deliberately caught, though they will take a small baited hook.

Relationships • About eight drummers occur here; the others described are the drummer (No. 42), mado (No. 43), parore (No. 45), and blue maomao (No. 46).

Distribution • North Cape to East Cape, mainly around rocky headlands and offshore islands. Also in south-east Australia.

Recognition • Deepish body with a small head and mouth, which has several rows of small three-pointed teeth. *Colour:* body and fins bright blue, on occasions becoming dark grey, many golden spots in broken rows on the body.

Size • Average length 40–50 cm; maximum 75 cm.

Food • Seaweed and seafloor invertebrates.

Reproduction • Little-known, but free-spawning in shallow water producing pelagic eggs.

Capture • *Recreational fishers:* by line or setnet.

45 Parore

Girella tricuspidata

Family KYPHOSIDAE
Drummers, sea chubs

46 Blue maomao

Scorpis violaceus

Family KYPHOSIDAE
Drummers, sea chubs

Relationships • About eight drummers occur here; the others described are the drummer (No. 42), mado (No. 43), bluefish (No. 44), and blue maomao (No. 46).

Other Names • Blackfish, mangrove fish.

Distribution • North Island, mostly north of East Cape and Cape Egmont. A relatively common species of shallow, inshore rocky areas, weedbeds, mangrove swamps and similar habitats, generally less than 10 m. Also around south-eastern Australia.

Recognition • Prominent strong tail; small slightly downturned mouth with small teeth, which in males are three-pronged. *Colour:* silvery grey to pale olive green, with about 10 darker vertical bars on the side; lower head and belly pale yellow.

Size • Average length 30–40 cm; maximum about 45 cm.

Food • Mostly seaweed, but small crustaceans and marine worms are sometimes taken.

Reproduction • Little-known, but free-spawning in shallow water producing pelagic eggs.

Capture • *Recreational fishers:* usually by setnet, or speared, sometimes by line and small hook. *Commercially:* by setnet.

Relationships • About eight drummers occur here, including the very similar sweep; others described are the drummer (No. 42), mado (No. 43), bluefish (No. 44), and parore (No. 45).

Distribution • North Island, most common along the north-east coast, on rocky shores in clear water about headlands and offshore islands in depths to about 20 m. Present at Tasman Sea islands and perhaps rarely in south-east Australia, where the related sweep *S. lineolatus* occurs.

Recognition • Oval body, small head and small terminal mouth, clearly forked tail. *Colour:* bright blue; juveniles duller but with a yellow anal fin. (The sweep is more greyish-brown in colour, and its juveniles are greenish with orange spots but lack the yellow anal fin.)

Size • Average length 25–35 cm; maximum 40 cm.

Food • Midwater plankton, mainly crustaceans.

Reproduction • Little-known, but presumably free-spawning in shallow water producing pelagic eggs.

Capture • *Recreational fishers:* by line and small hook. *Commercially:* sometimes in setnets, occasionally purse seines. A popular species for divers to watch and photograph.

47 Longfinned boarfish

Zanclistius elevatus

Family PENTACEROTIDAE
Boarfishes, armourheads

48 Black angelfish

Parma alboscapularis

Family POMACENTRIDAE
Damselfishes

Relationships • Five boarfishes occur here; others described are (Book 2) the giant boarfish (No. 69) and southern boarfish (No. 70).

Distribution • Restricted to the North Island, most common in the north, mainly around our northern reefs, from 30 m to at least 60 m, but also occurs on quite open seafloor down to about 200 m. The commonest of several species of boarfish in New Zealand. Also southern Australia.

Recognition • A distinctively shaped fish, well described by its common name. Narrow, highly arched body with a tall dorsal fin, and small protruding snout. *Colour:* silvery grey, sometimes faintly greenish, with indistinct oblique brown bands or blotches, plus a prominent dark spot near the rear of the dorsal fin.

Size • Average length 20–30 cm; maximum 35 cm.

Food • Small invertebrates, picked out from rocky crevices and encrusting seafloor growth by means of the pointed snout.

Reproduction • Unknown; possibly paired spawning near reefs.

Capture • *Recreational fishers:* accidentally in setnets, sometimes speared. *Commercially:* accidentally by trawl and setnet.

Relationships • Nine damselfishes occur here, mostly restricted to the far north; the other common species described is the demoiselle (No. 49).

Distribution • North Cape to about Hawke Bay, on open rocky shores to about 10 m. A similar species occurs in Australia, and both are present at Tasman Sea islands.

Recognition • Small head and almost oval body, pointed second dorsal and anal fins, large scales. *Colour:* dark grey to black. A white shoulder blotch, behind the top of the gill cover, most prominent in breeding adults (and which can quickly fade and reappear). Juveniles are yellowish brown with broken rows of small bright blue spots, blue fin margins, and a dark blotch just below the start of the soft dorsal fin.

Size • Average length 20–25 cm; maximum about 30 cm.

Food • Mainly fine seaweeds; small invertebrates are probably eaten accidentally.

Reproduction • The male prepares a nest on a vertical rock face by removing encrusting animals and cropping algae to a short turf; he fertilises eggs laid by several females, and guards them from predatory fishes.

Capture • Occasionally speared, mostly observed and photographed.

49 Demoiselle
Chromis dispilus

Family POMACENTRIDAE
Damselfishes

Relationships • Nine damselfishes occur here, mostly restricted to the far north; the other common species described is the black angelfish (No. 48). The single-spot demoiselle *C. hypsilepsis* and yellow demoiselle *C. fumea* are less often seen.

---

50 Kelpfish/Hiwihiwi/Ngakoikoi
Chironemus marmoratus

Family CHIRONEMIDAE
Kelpfishes

Distribution • Northern North Island, mainly the north-east coast. One of New Zealand's best-known, most abundant and attractive schooling reef fishes. Present along rocky coasts and around offshore islands from the surface down to about 50 m. Appears to be restricted to New Zealand, but similar species occur in Australia and elsewhere.

Recognition • Deep-bodied but narrow, with a pointed head and deeply forked tail. *Colour:* blue-grey, slightly paler below, but may vary from a light grey to almost black. Two conspicuous white spots on the back, just behind midbody and near the tail. Juveniles sometimes greenish. (The single-spot demoiselle is yellowish green with only one spot on each side near the tail base; the yellow demoiselle is yellow above, silvery white below, with two brown tail stripes.)

Size • Average length 10–20 cm.

Food • Midwater feeders on plankton, mainly crustaceans.

Reproduction • Males establish and guard nest sites on rock surfaces, in which several females deposit eggs to be fertilised; females may lay eggs in several nests.

Capture • *Recreational fishers:* by line and small hook, accidentally in setnets. *Commercially:* occasionally in setnets. A popular species for divers to watch and photograph.

Relationships • Two kelpfishes occur here.

Distribution • New Zealand's warm northern waters, less common from East Cape to Cook Strait. Also in southern Australia. Normal habitat is coastal reefs, in caves or crevices, particularly where there is dense seaweed and turbulent water, from the shore to about 20 m. (The second species occurs only at Kermadec Is.)

Recognition • Body roughly triangular in cross-section, sloping and pointed head with a relatively large mouth; large pectoral fins, which with the ventral fins (pelvic, anal) are used as supports to maintain position in strong currents. *Colour:* olive-brown above, mottled with green and red, with numerous small white spots; yellowish below.

Size • Average length 25–30 cm; maximum 35 cm.

Food • Crustaceans, shellfish, and small fishes.

Reproduction • Unknown.

Capture • *Recreational fishers:* in setnets, on lines, and by spear. *Commercially:* sometimes in shallow setnets.

51 Marblefish/Kehe

Aplodactylus arctidens

Family APLODACTYLIDAE
Marblefishes

52 Porae

Nemadactylus douglasi

Family CHEILODACTYLIDAE
Morwongs, fingerfins

Relationships • Two species occur here.

Other Names • Marble trout, granite trout.

Distribution • Inhabits shallow rock and weed-strewn areas throughout New Zealand waters, to about 20 m. More abundant in the southern half of its range. Also in cool Australian seas. (The notch-head marblefish *A. etheridgii* is common at the Kermadec Is and present but rare at some north-east New Zealand headlands.)

Recognition • Small rounded head, small downturned mouth with chisel-shaped teeth, plump body roughly triangular in cross-section. *Colour:* olive or brownish green, with white markings and spots. (*A. etheridgii* has a distinct notch behind the eyes, and numerous small white spots instead of irregular short lines.)

Size • Average length 30–50 cm; maximum 70 cm.

Food • Small seaweeds, mainly red turfing algae, grazed off rocks and (probably accidentally) some small animals from the weed.

Reproduction • Unknown.

Capture • *Recreational fishers:* in setnets or by spear. *Commercially:* accidentally in setnets. Seldom deliberately sought, the flesh having an unpleasant smell.

Relationships • Eight morwongs occur here; others described are the tarakihi (No. 53) and red moki (No. 54).

Distribution • North Island and northern South Island, but uncommon south of East Cape. Occurs over sandy areas of seafloor, usually in the vicinity of reefs, from shallow water down to about 100 m. Also in eastern Australia.

Recognition • Similar to the well-known tarakihi, but with thicker lips, a straighter head profile and smoothly arched back. No dark saddle mark behind the head. *Colour:* bluish or greenish grey above, often with a golden sheen, silvery below.

Size • Average length 30–45 cm; maximum 60 cm.

Food • Various invertebrates, mainly those in or on the sandy seafloor; also small fishes.

Reproduction • Little-known, presumably free-spawning in coastal waters.

Capture • *Recreational fishers:* by line, setnet, and occasionally spear. *Commercially:* trawl and setnet. A good food fish.

53 Tarakihi

Nemadactylus macropterus

Family CHEILODACTYLIDAE
Morwongs, fingerfins

54 Red moki/Nanua

Cheilodactylus spectabilis

Family CHEILODACTYLIDAE
Morwongs, fingerfins

Relationships • Eight morwongs occur here; others described are the porae (No. 52) and red moki (No. 54).

Distribution • Widespread in New Zealand waters, the juveniles around coastal reefs, adults on sandy or muddy seafloor to depths of about 200 m. Also in cool Australian seas. (A second species, the king tarakihi, occurs uncommonly off northern New Zealand.)

Recognition • Small mouth, long pectoral fin rays, silvery body, dark mark behind the head. *Colour:* silvery back and sides, whitish below, with large fish developing a greyish tinge. A prominent black saddle marking across the back of the head. (The king tarakihi is larger, with a black-tipped pectoral fin.)

Size • Average length 30–45 cm; maximum 60 cm.

Food • Small seafloor invertebrates.

Reproduction • Free-spawning in coastal waters.

Capture • *Recreational fishers:* line and setnet. *Commercially:* trawl and setnet. A popular food fish.

Relationships • Eight morwongs occur here; others described are the porae (No. 52) and tarakihi (No. 53).

Distribution • North Island and northern South Island, more abundant north of East Cape, at reefs or over very rough bottom, from shallow water down to about 30 m. Also in south-eastern Australia.

Recognition • Smallish mouth with thick fleshy lips, prominent strong scales, and striking colour pattern. *Colour:* head brownish orange, body whitish with eight red-brown vertical bands, larger fish more uniformly red, scale margins red or orange.

Size • Average length 35–45 cm; maximum 60 cm.

Food • Various invertebrates on or in the sandy seafloor, or that live among seaweed.

Reproduction • Little-known; possibly paired spawning in shallow water, producing pelagic eggs.

Capture • *Recreational fishers:* usually in setnets or speared, sometimes by line. *Commercially:* usually in setnets. Because of diminishing numbers its capture is discouraged.

55 Blue moki/Moki

Latridopsis ciliaris

Family LATRIDIDAE
Moki, trumpeters

Relationships • Four members of this family occur here; the other described is the trumpeter (Book 2, No. 71).

---

56 Yelloweye mullet/Aua

Aldrichetta forsteri

Family MUGILIDAE
Mullets

Distribution • Widespread about the New Zealand coast but most abundant south of East Cape. Juveniles live in shallow, rocky regions, while adults generally school over the sandy seafloor in depths up to 150 m. Adults make annual migrations from the north-eastern South Island and Cook Strait to spawn near East Cape. Also in Tasmania. (The related copper moki *L. forsteri* is quite rare but has a similar distribution.)

Recognition • Similar to the tarakihi and porae in shape, with thick lips similar to the latter, but a smaller and more rounded pectoral fin. *Colour:* blue-grey, often with broad darker bands on the back, grading to silver-grey and white on the flanks and underside. (*L. fosteri* is slightly smaller, whitish with partly mottled brownish stripes along the back, and black edges on dorsal and tail fins.)

Size • Average length 55–70 cm; maximum at least 90 cm.

Food • Small invertebrates on or in the sandy seafloor.

Reproduction • Free-spawning in coastal waters.

Capture • *Recreational fishers:* usually in setnets or speared, occasionally by line. *Commercially:* by trawl or setnet. A popular food fish.

Relationships • Two mullets occur here; the other is the grey mullet (No. 57).

Other Names • Sprat or herring (wrongly), and common mullet.

Distribution • This is the common 'herring'-like fish of shallow coastal waters and river estuaries around New Zealand. Also occurs in cool Australian waters.

Recognition • The slender form, narrow head, pointed snout and absence of a fleshy surround to the eye distinguish this species from the generally larger grey mullet. *Colour:* pale bluish green back, silvery white sides and belly. Eye bright yellow.

Size • Average length 15–25 cm; maximum 45 cm.

Food • Grazes on the rich algal and organic detritus layer on the seafloor; also eats small crustaceans.

Reproduction • Free-spawning in shallow coastal seas.

Capture • *Recreational fishers:* by line, setnet, beach seine, and various forms of trap. Commonly caught by children fishing from wharves. *Commercially:* setnet and beach seine. Makes a good kipper-like product. Often used as food for fish, seabirds, and marine mammals in public aquaria and marinelands.

57 Grey mullet/Kanae

Mugil cephalus

Family MUGILIDAE
Mullets

58 Spotty/Paketi/Pakirikiri

Notolabrus celidotus

Family LABRIDAE
Wrasses

Relationships • About 20 wrasses occur here, mostly in the north, with some tropical species only reaching the Kermadecs. Others described are the scarlet wrasse (No. 59), banded wrasse (No. 60), Sandager's wrasse (No. 61), green wrasse (No. 62), orange wrasse (No. 63), combfish (No. 64), crimson cleaner (No. 65), and red pigfish (No. 66).

Relationships • Two mullets occur here; the other is the yelloweye mullet (No. 56).

Distribution • Most common around the northern North Island coasts; it extends to northern South Island waters, but is much less common in the southern part of its range. Normal habitat is in shallow muddy harbours, mangrove swamps and river estuaries; also ascends rivers into purely fresh water. Widely distributed in warm seas throughout the world.

Recognition • The stout form, grey back, broad flat head and a fleshy surround and covering to the eye distinguish this species from the yelloweye mullet. *Colour:* greyish back, pale greyish to silver, faintly striped sides. Eye dull yellow.

Size • Average length 30–40 cm; maximum 60 cm.

Food • Feeds among the rich organic detritus and algae of the seafloor, also on mud-dwelling crustaceans and other animals.

Reproduction • Free-spawning in shallow coastal seas.

Capture • *Recreational fishers:* setnets and beach seines. *Commercially:* mainly in large setnets, often set around a school in shallow harbours, or drifted in large river estuaries. A popular food fish, fresh or smoked.

Other Names • Kelpie, guffy, sometimes butterfish (misleading).

Distribution • Widespread from North Cape to Stewart Island about shallow rocky places, mainly sheltered coasts and harbours. Restricted to New Zealand.

Recognition • 'Wrasse-like', with no distinctive features, though all but the largest males are quickly recognised by most people. *Colour:* variable: essentially brownish-olive to yellowish, but varying with its surroundings. Small fish (mostly females) to about 20 cm, have a distinct blotch and four darker vertical bars; in larger fish (mostly males) the blotch becomes a series of smaller spots, there are only two vertical bars, the body and fin colouring is brighter, with some light blue lines on the head.

Size • Average length 15–25 cm; maximum 27 cm.

Food • Bottom-dwelling invertebrates. Juveniles sometimes act as cleaner fish, picking skin parasites from other fish species.

Reproduction • Females change sex to males at about 20 cm. Spawning occurs in pairs in shallow water, the eggs are planktonic.

Capture • *Recreational fishers:* taken on light lines or in traps from around wharf piles or shallow rocks. Frequently caught by children.

59 Scarlet wrasse/Puuwaiwhakarua

Pseudolabrus miles

Family LABRIDAE
Wrasses

Relationships • About 20 wrasses occur here, mostly in the north, with some tropical species only reaching the Kermadecs. Others described are the spotty (No. 58), banded wrasse (No. 60), Sandager's wrasse (No. 61), green wrasse (No. 62), orange wrasse (No. 63), combfish (No. 64), crimson

---

60 Banded wrasse

Notolabrus fucicola

Family LABRIDAE
Wrasses

Relationships • About 20 wrasses occur here, mostly in the north, with some tropical species only reaching the Kermadecs. Others described are the spotty (No. 58), scarlet wrasse (No. 59), Sandager's wrasse (No. 61), green wrasse (No. 62), orange wrasse (No. 63), combfish (No. 64), crimson

cleaner (No. 65), and red pigfish (No. 66).

Other Names • Soldier, red soldier.

Distribution • Widespread on rocky ground between 15 m and about 100 m (i.e. deeper than most other common wrasses); more abundant south of the Bay of Plenty. Restricted to New Zealand.

Recognition • A typical wrasse in shape, with protruding teeth. The tail tips become pointed in adults, particularly mature males. *Colour:* pinkish red head and upper body, yellowish sides and belly, patterned with red; a black bar at the base of the tail. There are three size/sex phases: juveniles are pale pink and yellow, females have yellow sides with pink stripes, and males have yellowish orange sides with the rear edge of each scale bright red.

Size • Average length 20–30 cm; maximum 35 cm.

Food • A wide variety of bottom-dwelling invertebrates.

Reproduction • Females change sex to males at 15–20 cm. Spawning occurs in pairs in shallow water, the eggs becoming planktonic.

Capture • *Recreational fishers:* by line over rough ground, often a nuisance to anglers fishing for blue cod or tarakihi.

---

cleaner (No. 65), and red pigfish (No. 66).

Other Names • Kelpie, kelpfish.

Distribution • Widely distributed along rocky shores. Most abundant in central and southern New Zealand; lives in shallow water among rock and weed down to about 30 m. Also around southern Australia.

Recognition • Deep-bodied, with prominent teeth and very large rounded scales. *Colour:* variable, depending on age, sex, breeding condition, and the surrounding weed cover. Juveniles are mottled brownish or greenish with 5 pale spots on the back and fin. Adults brownish green to greyish purple, with darker bands on the side as well as the series of yellowish back and fin spots.

Size • Average length 25–35 cm; maximum 45 cm.

Food • Mainly crustaceans and shellfish, but also other small invertebrates.

Reproduction • Not well known; there is usually a change of sex from female to male, but it seems to occur over a range of sizes, and is not consistent. Paired spawning produces pelagic eggs.

Capture • *Recreational fishers:* in setnets, speared, or occasionally by line and small hook. *Commercially:* accidentally in setnets.

61 Sandager's wrasse

Coris sandageri

Family LABRIDAE
Wrasses

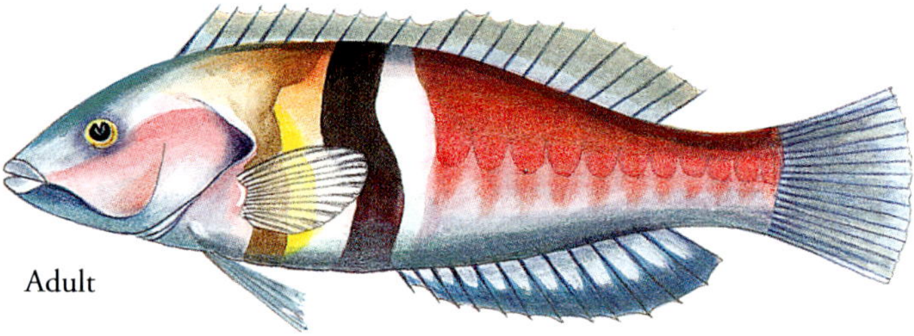

Adult

Juvenile

Relationships • About 20 wrasses occur here, mostly in the north, with some tropical species only reaching the Kermadecs. Others described are the spotty (no. 58), scarlet wrasse (No. 59), banded wrasse (No. 60), green

---

62 Green wrasse

Pseudolabrus inscriptus

Family LABRIDAE
Wrasses

Relationships • About 20 wrasses occur here, mostly in the north, with some tropical species only reaching the Kermadecs. Others described are the

wrasse (No. 62), orange wrasse (No. 63), combfish (No. 64), crimson cleaner (No. 65), and red pigfish (No. 66).

Distribution • North Cape to East Cape, around islands, headlands, and reefs, near the boundary of rock and sand, down to about 50 m. Also in south-eastern Australia.

Recognition • Typically wrasse-like, slender when young, deeper-bodied when adult. *Colour:* varies greatly with size and sex. Juveniles: whitish with a broad yellow stripe starting at the eye and ending at a large black dot at the tail base. Adult females: pinkish brown, with two brown blotches on the side. Males develop a series of colourful bands encircling the midbody, colour patches on the head, and bright fin patterns.

Size • Females reach about 35 cm in length; males are 35–45 cm.

Food • Juveniles often behave as 'cleaners', picking parasites from larger fish of the same and other species; they also eat small seafloor invertebrates as do the adult fish.

Reproduction • The typical wrasse sex change from female to male occurs at about 35 cm; paired spawning produces pelagic eggs.

Capture • Tame, easily observed and photographed by divers and quickly attracted by hand-feeding. Seldom deliberately caught, but will take a small baited hook.

---

spotty (No. 58), scarlet wrasse (No. 59), banded wrasse (No. 60), Sandager's wrasse (No. 61), orange wrasse (No. 63), combfish (No. 64), crimson cleaner (No. 65), and red pigfish (No. 66).

Distribution • North Cape to the Bay of Plenty, at islands and headlands, at or near reefs. Seen in New Zealand only since the late 1960s, probably having colonised from the Tasman Sea islands to the north-west, where it has a similar distribution to the orange wrasse. Local breeding may not occur, and the population size fluctuates.

Recognition • A large, attractively patterned wrasse. *Colour:* juveniles are brownish, with a green-tinged head and pale horizontal striping. Females remain essentially brown with intricate yellowish scale markings, while the large males are dark green with yellow dots on the scales.

Size • Average length 30–60 cm, depending on sex.

Food • A variety of seafloor invertebrates; large males also eat small fish such as triplefins.

Reproduction • A sex change from female to male at about 30 cm; paired spawning produces pelagic eggs.

Capture • Observed and photographed by divers; seldom caught.

63 Orange wrasse

Pseudolabrus luculentus

Family LABRIDAE
Wrasses

Relationships • About 20 wrasses occur here, mostly in the north, with some tropical species only reaching the

---

64 Combfish

Coris picta

Family LABRIDAE
Wrasses

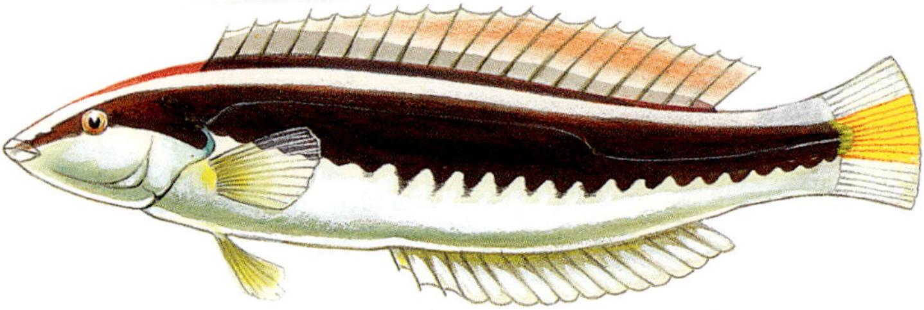

Relationships • About 20 wrasses occur here, mostly in the north, with some tropical species only reaching the Kermadecs. Others described are the

Kermadecs. Others described are the spotty (No. 58), scarlet wrasse (No. 59), banded wrasse (No. 60), Sandager's wrasse (No. 61), green wrasse (No. 62), combfish (No. 64), crimson cleaner (No. 65), and red pigfish (No. 66).

Distribution • North Cape to the Bay of Plenty, at islands and headlands, at or near reefs. Also in south-eastern Australia and at Tasman Sea islands.

Recognition • Typically wrasse-like. Adult males develop a blunter head and slightly deeper body. *Colour:* juveniles and females are reddish overall, with pale lines on the lower head continued as white scale markings on the flanks. Males are greenish around the head and sometimes the forward flanks, with distinct black and white markings at the rear of the dorsal fin base, and a blue spot at the pectoral fin base.

Size • Average length 15–25 cm, depending on sex.

Food • Crustaceans. Juveniles also behave as cleaner fish.

Reproduction • A sex change from female to male at 15–20 cm; paired spawning produces pelagic eggs.

Capture • Observed and photographed by divers; seldom caught.

---

spotty (No. 58), scarlet wrasse (No. 59), banded wrasse (No. 60), Sandager's wrasse (No. 61), green wrasse (No. 62), orange wrasse (No. 63), combfish (No. 64), crimson cleaner (No. 65), and red pigfish (No. 66).

Distribution • North Cape to East Cape, best known at the Poor Knights Islands, but occurs at several islands and headlands on the north-east coast. A presumed immigrant in the late 1960s, it has fluctuated in numbers since then. Also in eastern Australia.

Recognition • Small, slender. *Colour:* strikingly marked, mostly white with a broad black stripe from snout to tail and a thin reddish stripe at the dorsal fin base. In adults the tail portion of the black stripe becomes yellow; spawning males briefly adopt a more complex pattern of different coloured stripes and patches.

Size • Average length 10–20 cm, the largest fish being males.

Food • Small crustaceans. Many combfish adopt an almost exclusively cleaning behaviour (see Crimson cleaner, No. 65) on larger fishes.

Reproduction • Pattern of sex change not well known; paired spawning produces pelagic eggs.

Capture • Observed and photographed by divers; seldom caught.

65 Crimson cleaner

Suezichthys aylingi

Family LABRIDAE
Wrasses

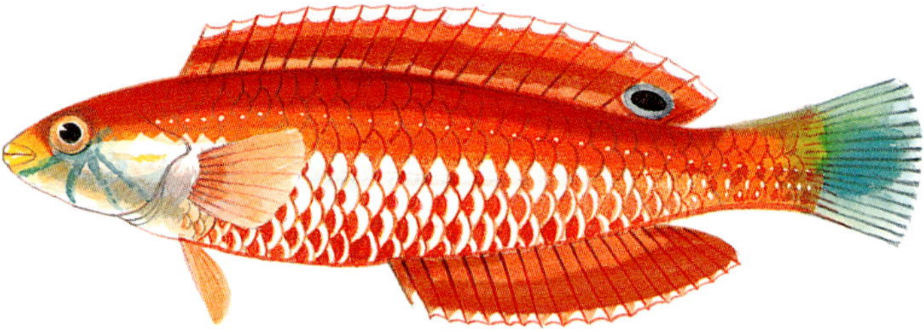

Relationships • About 20 wrasses occur here, mostly in the north, with some tropical species only reaching the

66 Red pigfish/Pakurakura

Bodianus vulpinus

Family LABRIDAE
Wrasses

Relationships • About 20 wrasses occur here, mostly in the north, with some tropical species only reaching the

Kermadecs. Others described are the spotty (No. 58), scarlet wrasse (No. 59), banded wrasse (No. 60), Sandager's wrasse (No. 61), green wrasse (No. 62), orange wrasse (No. 63), combfish (No. 64), and red pigfish (No. 66).

Distribution • North Cape to East Cape, mainly at islands and headlands, only locally common. Also in south-eastern Australia.

Recognition • Small, slender. *Colour:* striking – mainly red and yellowish orange, with blue and white markings. Females: reddish orange with white scale markings on the flanks and a black spot on the dorsal fin. Males: darker red with a white line along each side and blue lines on the head and on each scale.

Size • Average length 10–15 cm, the larger fish males with a small harem of females.

Food • Has a strongly developed 'cleaning' behaviour, taking skin parasites (crustaceans) from other fishes. However, this is variable, some fish obtaining almost all their food this way, and others doing no cleaning at all.

Reproduction • A sex change from female to male at 10 cm; paired spawning produces pelagic eggs.

Capture • Observed and photographed by divers; seldom caught.

Kermadecs. Others described are the spotty (No. 58), scarlet wrasse (No. 59), banded wrasse (No. 60), Sandager's wrasse (No. 61), green wrasse (No. 62), orange wrasse (No. 63), combfish (No. 64), and crimson cleaner (No. 65).

Other Names • Spotted pigfish. Previously listed as *Verreo oxycephalus.*

Distribution • North Cape to East Cape, at reefs and over the rocky seafloor, from shallow water to about 50 m. Widely distributed in the Pacific, between Australia, Japan and Easter Island.

Recognition • A typical wrasse in body shape, but with a long snout and large mouth. Large males have extended tail tips. *Colour:* females are yellowish pink with a series of thin red lines along the side, plus three rows of elongated red blotches; males brighter red with fainter lines, a pale yellow blotch on the back at the rear of the soft dorsal fin, and a large bluish black spot on the spiny dorsal fin.

Size • Average length 25–35 cm; maximum 45 cm.

Food • A variety of seafloor invertebrates, including hard-shelled crustaceans and shellfish.

Reproduction • A sex change from female to male at 25–30 cm; paired spawning produces pelagic eggs.

Capture • *Recreational fishers:* by line or setnet near rough ground.

67 Butterfish/Marari/Rari

Odax pullus

Family ODACIDAE
Butterfishes

Relationships • Two butterfishes occur here.

Other Names • The name greenbone, often used in southern New Zealand, refers to the bone colour.

---

68 Thornfish

Bovichtus variegatus

Family BOVICHTHYIDAE
Thornfishes

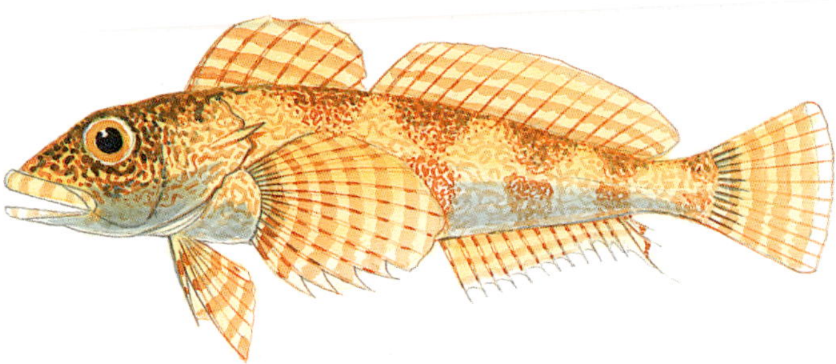

Relationships • Three thornfishes occur here.

Distribution • Widespread in shallow rock and weedbed areas around the coasts, to about 20 m. Much more abundant south of East Cape. Restricted to New Zealand. (Blue-finned butterfish *O. cyanoallix* occurs at the Three Kings Is.)

Recognition • Slender body, pointed head with a small mouth, teeth fused into a small beak, large dorsal and anal fins, particularly in mature males. *Colour:* variable; juveniles are golden yellow with a broken white midbody stripe; adults become brownish to olive green, the males with light blue markings. (*O. cyanoallix* is similar, smaller [to 35 cm] but deeper-bodied, and has slightly different colour phases, mature males being distinguished by blue-fringed fins.)

Size • Average length 35–45 cm; maximum 60 cm.

Food • Mainly kelp and other seaweeds, but also small crustaceans and other invertebrates.

Reproduction • Small fish are females, changing sex to males at 35–40 cm. Males establish territories, pairing with several females during the spawning season; larvae settle from planktonic eggs.

Capture • *Recreational fishers:* by setnet or spear, occasionally a very small hook. *Commercially:* by setnet. A popular food fish.

Other Names • Sometimes called horny, from its rather spiny appearance, or (erroneously) dragonet, from a general resemblance to the colourful, large-finned fishes of an unrelated family.

Distribution • Uncommon around the southern North Island, more common from Cook Strait southwards to New Zealand's subantarctic islands, from tidepools to about 50 m. Although colourful and conspicuous when seen alone, it is well camouflaged for its subtidal rocky and weedy habitat. (A second species occurs only at the subantarctic islands, and a third in deep water also in the south. Another species of thornfish occurs around southern Australia.)

Recognition • A broad, sloping head, large eyes, a prominent spine on the gill cover, and a single lateral line. *Colour:* variable, but essentially brown or grey, mottled and striped (particularly on the fins) with green, orange, pink, and white. Juveniles: paler with brownish vertical bands.

Size • Average length 15–20 cm; maximum 25 cm.

Food • Seafloor invertebrates, mainly crustaceans.

Reproduction • Unknown.

Capture • Sometimes taken on a small hook, more often seen underwater by divers, or darting across a deep tidepool.

69 Spotted stargazer/Kourepoua

Geniagnus monopterygius

Family URANOSCOPIDAE
Armourhead stargazers

70 Estuary stargazer

Leptoscopus macropygus

Family LEPTOSCOPIDAE
Stargazers

Relationships • There are five stargazers from this family here (the other one described is the monkfish [Book 2, No. 75]), and three from a related stargazer family (see the estuary stargazer [No. 70]).

Distribution • Widespread in New Zealand waters on a sandy to muddy seafloor near the shoreline and out to depths of 100 m. Restricted to New Zealand.

Recognition • A moderately plump stargazer-shaped body, lacks the strong spines of the giant stargazer, and has a small chin barbel. Most easily recognised by the colour pattern. *Colour:* dark greyish-green or brownish above, with large oval or circular cream spots; whitish below.

Size • Average length 25–35 cm; maximum 45 cm.

Food • Buries itself in soft sediments, leaving only the eyes and upturned mouth exposed. Any suitable prey animals passing over or near the mouth are taken in a swift gulp. Crustaceans and small fish are taken this way, and shellfish located in the sediment also eaten.

Reproduction • Unknown; presumably free-spawning in coastal waters.

Capture • *Recreational fishers:* by line, beach seine, or flounder spear. *Commercially:* usually by trawl. A good food fish.

Relationships • There are three stargazers from this family here, the other two being less common. There are also five species in a related stargazer family; others described are the spotted stargazer (No. 69) and monkfish (Book 2, No. 75).

Distribution • Throughout New Zealand's shallow coastal waters and estuaries, including the lower reaches of large rivers, to about 20 m. Also in south-eastern Australia. (*Crapatalus novaezelandiae* and *C. angusticeps* occupy the same shallow habitat as the estuary stargazer.)

Recognition • A typical slender stargazer in shape. *Colour:* greyish above with many small black spots and irregular blotches, a whitish band along each side with a dark band below it; lower body and belly pinkish white. (*C. novaezelandiae* is also speckled and white but lacks the side bands. *C. angusticeps* is similarly coloured but much smaller and has a narrow head.)

Size • Average length 20–30 cm; maximum 35 cm.

Food • Bottom-living invertebrates and small fishes.

Reproduction • Unknown, but presumably free-spawning in shallow coastal waters.

Capture • *Recreational fishers:* sometimes caught in flounder nets. Usually buried in the sand, and disturbed by swimmers and waders.

71 Opalfish

Hemerocoetes monopterygius

Family PERCOPHIDAE
Opalfishes, duckbills

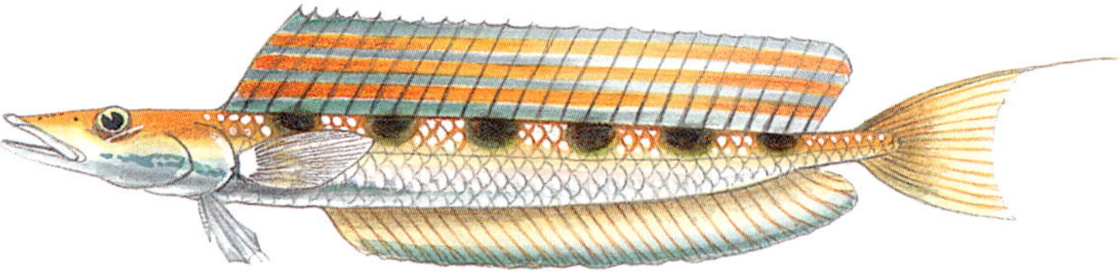

72 Blue cod/Raawaru/Pakirikiri, Patutuki

Parapercis colias

Family PINGUIPEDIDAE
Sandperches

Relationships • Four members of this family occur here. Two are uncommon marine sandperches, the yellow cod (Book 2, No. 76) and the redbanded cod. The third is the small freshwater torrentfish.

Relationships • There are about eight species in this family here; this is the commonest.

Distribution • Though not frequently seen, the opalfish is a common and widely distributed species from shallow water out to 200 m. Restricted to New Zealand.

Recognition • Pointed snout, flattened head with almost upward-looking eyes, slender body and long, colourful dorsal fin. Mature males have an elongated upper tail tip. *Colour:* greyish with an iridescent pattern of green, blue and orange, with an opalescent sheen in live fish. (The related species are less common and not as colourful.)

Size • Average length 15–20 cm; maximum 25 cm.

Food • Crustaceans and small fishes.

Reproduction • Unknown.

Capture • *Recreational fishers:* sometimes caught on small hooks fished in muddy harbours, or tangled among debris in nets. *Commercially:* occasionally found among small fishes tangled in commercial trawls.

Distribution • Widespread in New Zealand waters but abundant only about and south of Cook Strait and at the Chatham Islands. Normally lives on rocky ground from the shallows to depths of about 150 m. Restricted to New Zealand, but many related sandperches occur in the Indo-Pacific region.

Recognition • Plump body shape, sloping and prominent head, large eyes, small first dorsal fin. *Colour:* juveniles are whitish with two dark stripes along the back, these stripes broaden and become mid brown in medium sized fish; the largest fish (males) are blue-grey with a greenish sheen. (The two other sandperches are coloured as their common names suggest.)

Size • Average length 30–40 cm; maximum 60 cm.

Food • Wide variety of small and large seafloor invertebrates, also smallish fishes. Can wrench limpets and young paua off rocks with its small but strong teeth.

Reproduction • At least some females change sex to males; free-spawning in coastal waters.

Capture • *Recreational fishers:* almost exclusively line-caught. *Commercially:* by line, also in cod-pots, and sometimes trawled near rough ground. A popular food fish.

Rockpool fishes

Many small fishes inhabit rock pools. Some are quite distinctive, but others (particularly those known as 'cockabullies' and 'blennies') are difficult to identify. The latter belong to several families: the triplefins (Nos. 73–77), a weedfish or clinid (No. 78), a true blenny (No. 79), and a sleeper or gudgeon (No. 80). Also found in tidepools are clingfishes (No. 21), pipefishes (Nos. 25, 26), the seahorse (No. 27), and the olive rockfish (No. 34).

Scalyheaded triplefin

Variable triplefin

73 Scalyheaded triplefin

Karalepis stewarti

Family TRIPTERYGIIDAE
Triplefins, threefin blennies

Relationships • Of the 30 or so triplefins occurring here about 20 are reasonably common, including this species, the variable triplefin (No. 74) longfinned triplefin (No. 75), twister (No. 76), and topknot (No. 77).

Distribution • Widespread around New Zealand, generally more common in the North Island, among rocks rather than in tidepools. Restricted to New Zealand.

Recognition • Secretive and nocturnal in behaviour. Distinguished from related New Zealand species by a dense and rough scaly covering on the head. *Colour:* banded and mottled reddish brown, effective camouflage for its rocky habitat.

Size • Average length 10 cm; maximum about 15 cm.

Food • Small bottom-living crustaceans and shellfish.

Reproduction • Unknown.

Capture • Can be caught by hand.

74 Variable triplefin

Forsterygion varium

Family TRIPTERYGIIDAE
Triplefins, threefin blennies

Relationships • Of the 30 or so triplefins occurring here about 20 are reasonably common, including this species, the scalyheaded triplefin (No. 73), the longfinned triplefin (No. 75), twister (No. 76), and topknot (No. 77).

Distribution • Common and widespread, among rocks and weeds about and below low-tide mark, to about 30 m. Also in Tasmania.

Recognition • A generalised triplefin. *Colour:* variable: brownish to greenish grey above, about 6 dark bands above the lateral line separated by light markings on the back, lighter and specked below; head mottled reddish brown. Breeding males darker, with white fin margins.

Size • Average length 8–10 cm; maximum 13 cm.

Food • Bottom-living crustaceans and other invertebrates.

Reproduction • Several females lay eggs in a sheet on the side of a rock, where they are guarded by a male until they hatch.

Capture • Can be caught by hand, and will occasionally take a very small hook.

75 Longfinned triplefin

Ruanoho decemdigitatus

Family TRIPTERYGIIDAE
Triplefins, threefin blennies

76 Twister

Bellapiscis medius

Family TRIPTERYGIIDAE
Triplefins, threefin blennies

Relationships • Of the 30 or so triplefins occurring here about 20 are reasonably common, including this species, the scalyheaded triplefin (No. 73), variable triplefin (No. 74), twister (No. 76), and topknot (No. 77).

Distribution • Widespread in sheltered rocky intertidal and subtidal habitats, to about 6 m. Restricted to New Zealand.

Recognition • Snout pointed, head flattened, with a fan-like fringe above the eye, scales relatively large. *Colour:* males dark grey with reddish fins, females mottled greenish grey.

Size • Average length 6–8 cm; maximum about 10 cm.

Food • Little-known, but presumably small crustaceans and other invertebrates.

Reproduction • Egg masses are laid beneath stones by several females, and guarded by a male until they hatch.

Capture • Can be caught by hand in low-tidal pools.

Relationships • Of the 30 or so triplefins occurring here about 20 are reasonably common, including this species, the scalyheaded triplefin (No. 73), variable triplefin (No. 74), longfinned triplefin (No 75), and topknot (No. 77).

Distribution • A common species of rock pools, the twister is widespread wherever a suitable habitat occurs. Restricted to New Zealand.

Recognition • A low first dorsal fin of four spines, a short steep snout, and no tentacle over the eye. However, other triplefins are very similar, and have formerly been confused with this species. *Colour:* variable: greyish to brown, with checkerboard markings, mainly below the lateral line.

Size • Average length 5–7 cm; maximum about 8 cm.

Food • Small bottom-living crustaceans and shellfish.

Reproduction • Eggs are laid in a sheet on a rock's undersurface, and fertilised and guarded by the male until they hatch.

Capture • By hand (scoop or dip net) from rock pools. An interesting native fish for marine aquaria.

77 Topknot/Hetarua

Notoclinus fenestratus

Family TRIPTERYGIIDAE
Triplefins, threefin blennies

78 Crested weedfish

Cristiceps aurantiacus

Family CLINIDAE
Clinids, weedfishes

Relationships • Three weedfishes occur here.

Relationships • Of the 30 or so triplefins occurring here about 20 are reasonably common, including this species, the scalyheaded triplefin (No. 73), variable triplefin (No. 74), longfinned triplefin (No 75), and twister (No. 76).

Distribution • Widespread in shallow water, though infrequently seen because it is well camouflaged for its habitat of brown seaweed. (The smaller brown topknot *N. compressus* occurs in northern and central New Zealand.) Both species crawl over the seaweed using their ventral fins.

Recognition • Body narrow rather than rounded as in other triplefins, first dorsal fin high and set well forward on the head, hence 'topknot', transparent areas on fins. *Colour:* variable, but usually reddish brown or orange with darker bars and mottling. (The brown topknot has a smaller mouth, reaching to below the front [not the rear] of the eye, and its second [not first] dorsal fin spine is longest. Note: the weedfishes [No. 78] also have clear fin areas.)

Size • Average length 5–10 cm; maximum about 20 cm.

Food • Small weed-dwelling crustaceans.

Reproduction • Unknown.

Capture • Can be caught by hand.

Distribution • North-east New Zealand, to about Hawke Bay. Inhabits shallow beds of seaweed and seagrass, from the shoreline out to about 30 m. Also southern Australia. (The smaller orange clinid *Ericentrus rubrus* extends to central New Zealand in the same habitat, and there is a third widespread but rare species.)

Recognition • Blunt-headed, body narrow and rather tapering, with a thin tail base and small rounded tail. The short, high first dorsal fin is above the eye, hence 'crested'. *Colour:* variable: a blotched and banded pale orange to dark brown, sometimes greenish, with the fins alternately patterned and translucent like the topknots (No. 77), camouflaging all these fishes well for an inconspicuous life among the weed. (The orange clinid, also variable in colour, has its small first dorsal fin attached to the second.)

Size • Average 15–20 cm; maximum about 25 cm.

Food • Small weed-dwelling crustaceans and shellfish, also small fishes.

Reproduction • Gives birth to up to several hundred live larvae instead of simply releasing eggs to be externally fertilised.

Capture • Sometimes caught in nets or lobster pots. Being well camouflaged, weedfishes are only seen by observant divers.

79 Crested blenny

Parablennius laticlavius

Family BLENNIIDAE
Blennies

Relationships • Two blennies occur here, plus at least four at the Kermadecs.

Distribution • Most common in the northern North Island, from rockpools down to about 25 m. They

---

80 Graham's gudgeon/Kurahina

Grahamichthys radiata

Family ELEOTRIDIDAE
Bullies, sleepers

occur near reefs or on a rough sea-floor, where they can shelter in small holes. Restricted to New Zealand.

Recognition • One of the most easily recognised of the many kinds of small fishes seen in tidepools and adjacent shallow water. A blunt head and narrow, tapering body. The dorsal and anal fins are long, and the head has a forked frond (cirrus) above each eye, hence 'crested'. *Colour:* pale yellowish, with a broad dark band running from eye to tail base, often with a silvery stripe below this; the head has smaller stripes, and may be darkish in large fish. (The other blenny, the mimic blenny *Plagiotremus tapeinosoma*, is very slender, dark green and yellowish with darker bands and silver stripes; it mimics harmless triplefins by swimming with them until it can approach an unsuspecting larger fish and nip off a piece of skin or fin.)

Size • Average length 3–5 cm; maximum 8 cm.

Food • Small bottom-living crustaceans, fish and mollusc eggs, newly settled or small encrusting invertebrates, and algae.

Reproduction • Eggs are laid in a seafloor crevice and guarded until they hatch.

Capture • Can be caught by hand. More often observed and photographed by divers.

Relationships • Six members of this family occur in our estuarine and fresh waters; this is our only marine bully.

Distribution • Widely distributed in shallow water in sheltered bays, usually where rocks or shells provide cover on an otherwise open sandy or muddy seafloor. Common in some localities. Restricted to New Zealand.

Recognition • A goby-like fish, with broad head and mouth, tapering only slightly towards the tail. Large pectoral fins, the pelvics small and not fused together like those of most gobies. Distinguished from the triplefins by having only two dorsal fins. *Colour:* pale greyish brown, quite translucent when small, and banded with darker brown when adult.

Size • Average length 3–5 cm; maximum 6 cm.

Food • Unknown; presumably small invertebrates.

Reproduction • Unknown.

Capture • In small-meshed nets, when the transparent juveniles may be mistaken for whitebait.

81 Barracouta/Manga/Maka

Thyrsites atun

Family GEMPYLIDAE
Snake mackerels

82 Blue warehou

Seriolella brama

Family CENTROLOPHIDAE
Medusafishes, warehous

Relationships • Nine members of this family occur here but most are uncommon or rare; the other described is the gemfish (Book 2, No. 77).

Distribution • Widespread around New Zealand, ranging from near the shoreline out to the shelf edge at 200 m, and from the seafloor to the surface. Schooling and migratory, and seasonally abundant in many regions. Also in the cooler waters of South Africa, Australia and South America.

Recognition • Slender form, pointed head, large mouth and teeth, and single undulating lateral line. *Colour:* dark greyish blue on the top of the back, otherwise silvery with blue and purple iridescence.

Size • Average length 60–100 cm; maximum about 140 cm.

Food • Midwater plankton, particularly crustaceans, and small fishes.

Reproduction • Free-spawning in open coastal waters.

Capture • *Recreational fishers:* usually by line, especially driftline. *Commercially:* mainly by trawl.

Relationships • About a dozen members of this family occur here; others described are (Book 2) the silver warehou (No. 87), white warehou (No. 88), bluenose (No. 89), and rudderfish (No. 90).

Other Names • Common warehou.

Distribution • Widely distributed south of the Bay of Plenty, most common around and south of Cook Strait, from the shallows to about 100 m, usually on the open seafloor or in midwater above it, but sometimes near reefs. Also off southern Australia.

Recognition • Small head and mouth, rather deep-bodied and plump, long pointed pectoral fin reaching back to start of anal fin. *Colour:* bluish green above, slightly iridescent, silvery sides and belly. A prominent dark blotch behind the gill cover and above the pectoral fin base.

Size • Average length 50–60 cm; maximum about 75 cm.

Food • Some small seafloor invertebrates, but mainly midwater plankton, particularly jellyfish-like pelagic tunicates.

Reproduction • Free-spawning in open coastal waters.

Capture • *Recreational fishers:* mainly in setnets. *Commercially:* by trawl and setnet. A good food fish.

Flatfishes

The two families in New Zealand – left-eyed and right-eyed flounders – are named for the side of the body their eyes are on. Our best-known species are right-eyed flounders, including the sand flounder (No. 83), yellowbelly flounder (No. 84), greenback flounder (No. 85), black flounder (No. 86), turbot (No. 87), brill (No. 88), common sole (No. 89), and lemon sole (No. 90). There are three left-eyed flounders: the witch (Book 2, No. 91), crested flounder, and a large deepwater species.

Sand flounder

Yellowbelly flounder

83 Sand flounder/Patiki

Rhombosolea plebeia

Family PLEURONECTIDAE
Right-eyed flounders

Relationships • Eleven right-eyed flounders occur here; the eight most common species are described.

Other Names • Dab, diamond, tinplate and square flounder. The Maori name patiki is now used for most flounders.

Distribution • Widespread in inshore, mainly sandy areas; may also enter the mouths of large river estuaries. Restricted to New Zealand.

Recognition • A very wide, diamond-shaped body. *Colour:* dark greenish brown upper surface, white undersurface.

Size • Average length 25–35 cm; maximum 45 cm.

Food • A variety of small animals living in sediments, particularly brittlestars and marine worms, located by vision and touch.

Reproduction • Free-spawning in coastal seas.

Capture • *Recreational fishers:* usually in beach seines, setnets or by spear. *Commercially:* usually by trawl, often by setnet. A popular food fish.

84 Yellowbelly flounder

Rhombosolea leporina

Family PLEURONECTIDAE
Right-eyed flounders

Relationships • Eleven right-eyed flounders occur here; the eight most common species are described.

Distribution • Widespread in inshore areas, mainly on a sandy mud seafloor; common in sheltered bays, estuaries and harbours. Restricted to New Zealand.

Recognition • Oval body, not as angular as the sand flounder, and slightly plumper. *Colour:* greenish upper surface (greener than the sand flounder's), characteristic yellow (occasionally white) undersurface, often with scattered dark spots.

Size • Average length 25–35 cm; maximum 45 cm.

Food • Mainly burrowing seafloor invertebrates, particularly crabs and brittlestars, located by touch and taste.

Reproduction • Free-spawning in shallow coastal seas.

Capture • *Recreational fishers:* usually in beach seines, setnets or by spear. *Commercially:* usually by setnet, sometimes by trawl. A popular food fish.

85 Greenback flounder

Rhombosolea tapirina

Family PLEURONECTIDAE
Right-eyed flounders

---

86 Black flounder/Patiki mohoao

Rhombosolea retiaria

Family PLEURONECTIDAE
Right-eyed flounders

Relationships • Eleven right-eyed flounders occur here, and the eight most common species are described in this book.

Distribution • A moderately common flatfish of shallow coastal waters around the South Island. Also in southern Australia.

Recognition • Oval body, a triangular head and clearly pointed snout with a white fleshy tip to the upper jaw. *Colour:* dark green above, white below.

Size • Average length 25–40 cm; maximum 50 cm.

Food • A variety of mainly surface-living seafloor invertebrates, and small fishes.

Reproduction • Free-spawning in coastal seas.

Capture • *Recreational fishers:* usually in beach seines or setnets. *Commercially:* by trawl and setnet.

Relationships • Eleven right-eyed flounders occur here, and the eight most common species are described in this book.

Other Names • River flounder, mud flounder.

Distribution • One of the common flounders of river estuaries, particularly in central and southern regions. Sometimes travels many kilometres inland, where it can live in fresh water. It may also be found in the sea, but only near river mouths. Restricted to New Zealand.

Recognition • Broadly oval body. *Colour:* greenish brown, usually with distinctive red spots and blackish mottling on the upper surface; greyish undersurface, sometimes mottled.

Size • Average length 15–25 cm; maximum 45 cm.

Food • Much the same food as other flounders, and in rivers they take insects and small freshwater fishes such as whitebait.

Reproduction • Little known; probably free-spawning in shallow coastal seas.

Capture • *Recreational fishers:* setnets and spear. *Commercially:* setnets, sometimes in eel trap-nets.

87 **Turbot**
Colistium nudipinnis

Family PLEURONECTIDAE
Right-eyed flounders

88 **Brill**
Colistium guntheri

Family PLEURONECTIDAE
Right-eyed flounders

Relationships • Eleven right-eyed flounders occur here, and the eight most common species are described in this book.

Distribution • Coastal waters, principally around southern New Zealand, but extending in small numbers up both coasts of the North Island. Like the brill, the turbot is restricted to New Zealand, and not related to its namesake in Europe.

Recognition • Large, oval, rather broader than the brill, with a wide rounded head and the upper jaw extended as a hook or beak. *Colour:* brownish green above, with irregular dark and pale blotches; whitish with pale brown blotches below.

Size • Average length 25–45 cm; maximum at least 80 cm.

Food • Seafloor invertebrates, particularly crustaceans.

Reproduction • Free-spawning in coastal seas.

Capture • *Recreational fishers:* setnets and beach seines. *Commercially:* usually by trawl, sometimes setnet. A popular food fish.

Relationships • Eleven right-eyed flounders occur here, and the eight most common species are described in this book.

Distribution • Southern coastal waters, particularly around Otago. Restricted to New Zealand, and not related to the European brill.

Recognition • Large, broadly and almost symmetrically oval, with a rounded head and small mouth. *Colour:* brown to greenish in colour above, with fine black mottling in longitudinal lines; white and grey below, sometimes speckled.

Size • Average length 25–40 cm; maximum at least 70 cm.

Food • As might be expected for such a large flatfish, a wide variety of seafloor invertebrates as well as small fishes.

Reproduction • Free-spawning in coastal seas.

Capture • *Recreational fishers:* occasionally in setnets. *Commercially:* usually by trawl. A popular food fish.

89 Common sole

Peltorhamphus novaezeelandiae

Family PLEURONECTIDAE
Right-eyed flounders

90 Lemon sole

Pelotretis flavilatus

Family PLEURONECTIDAE
Right-eyed flounders

Relationships • Eleven right-eyed flounders occur here, and the eight most common species are described in this book. There are also two small relatives of the common sole.

Other Names • Sole, English sole or New Zealand sole.

Distribution • Widespread on inshore sandy areas, particularly bays, to depths of about 120 m, being most common from 10–20 m. Restricted to New Zealand.

Recognition • A broadly rounded snout with a fleshy overhang which obscures the mouth, and a lengthened first pectoral fin ray. *Colour:* greenish grey above, white below. (The two related species are much smaller [10–15 cm]; the speckled sole *P. latus* is similar in shape but has small dark spots, and the slender sole *P. tenuis* is narrower with faint dark lines.)

Size • Average length 25–35 cm; maximum 45 cm.

Food • Burrowing seafloor invertebrates, particularly marine worms, located by touch and taste.

Reproduction • Free-spawning in coastal seas.

Capture • *Recreational fishers:* usually in beach seines, setnets, or by spear. *Commercially:* by trawl or setnet. A popular food fish.

Relationships • Eleven right-eyed flounders occur here, and the eight most common species are described in this book.

Distribution • Moderately common in shallow to mid-shelf waters, most common in the south. Juveniles are found in sheltered areas, adults in open waters on a sandy or muddy bottom. Restricted to New Zealand, although superficially similar 'lemon soles' occur in various parts of the world.

Recognition • Somewhat oval in shape. The head is small and pointed, and the large eyes placed close to the projecting snout. The skin has distinct scales, in contrast to the smooth skin of other species. *Colour:* grey-brown, slightly greenish above, and white below.

Size • Average length 25–35 cm; maximum 50 cm.

Food • Worms and crustaceans living on the seafloor, located and stalked using the forward-placed eyes.

Reproduction • Free-spawning in the open sea.

Capture • *Recreational fishers:* like other flatfish they will sometimes take a small hook, but are usually beyond the depth range of other fishing methods. *Commercially:* usually by trawl. A popular food fish.

91 Leatherjacket/Kokiri

Parika scaber

Family MONACANTHIDAE
Leatherjackets

Relationships • About a dozen leatherjackets occur here, but only this species is common.

Other Names • Rough leatherjacket. Sold as creamfish. The names filefish and triggerfish are

---

92 Globefish

Contusus richei

Family TETRAODONTIDAE
Pufferfishes

sometimes used but they more properly apply to a related group of tropical fishes.

Distribution • Common about rocky shores throughout New Zealand, it also occurs over more open seafloors, where it ranges down to depths of about 50 m. Also around southern Australia. (The other leatherjackets occur at the Kermadecs or are only rare visitors to our northern waters.)

Recognition • Characteristic body shape, small mouth and exposed teeth, rough-textured skin, and strong dorsal spine above the eye. *Colour:* variable, from whitish with brown blotches through to a uniform greenish grey brown, gill slit blue, tail yellowish green in males with a vertical black line near the tip, brownish in females, other fins yellow.

Size • Average length 20–30 cm; maximum about 35 cm.

Food • Small encrusting and other seafloor invertebrates, also fine seaweed.

Reproduction • Paired spawning, with eggs deposited in undefended nests among rocks.

Capture • *Recreational fishers:* by line and small hook, setnet; being tame and inquisitive is easily speared. *Commercially:* usually by trawl. A good food fish.

Relationships • About nine pufferfishes occur here, but only this species is common.

Other Names • Puffer, toado.

Distribution • Widespread, but less common and more localised north of Cook Strait. Occurs off sandy beaches and in muddy bays and large estuaries, from the shore to about 50 m. Also in southern Australia. (The other, generally more colourful pufferfishes occur at the Kermadecs or are seen only rarely in our northern waters.)

Recognition • Small in size, with small mouth and teeth. Inflates itself to become almost spherical when disturbed. *Colour:* greyish brown with dark brown spots and blotches, white below.

Size • Average length 8–12 cm; maximum 15 cm.

Food • Small seafloor invertebrates, particularly crustaceans.

Reproduction • Unknown; presumably free-spawning in shallow coastal waters.

Capture • *Recreational fishers:* usually caught in beach seines, sometimes setnets; may also be taken on a small hook. This fish is poisonous.

93

Porcupinefish/Koputotara

Allomycterus jaculiferus

Family DIODONTIDAE
Porcupinefishes

Relationships • The only common New Zealand porcupinefish.

Distribution • Widespread and moderately common in coastal waters, generally over a sandy seafloor; may also enter harbours. More common north of Cook Strait. Lives from surface waters to depths of about 200 m, often in schools. Restricted to New Zealand, with a very similar species in Australia. (Another species occurs at the Kermadecs.)

Recognition • Blunt head, stout body with dorsal and anal fins near the tail base, strong backward-pointing spines. When disturbed it inflates with water and/or air and becomes almost spherical, with prominently projecting spines. *Colour:* pale brownish above, with large darker brown and black blotches surrounded by yellow spots and markings; white below.

Size • Average length 25–35 cm; maximum about 50 cm.

Food • Seafloor invertebrates, including hard-shelled crustaceans and shellfish.

Reproduction • Unknown; presumably free-spawning in coastal waters.

Capture • *Recreational fishers:* occasionally in beach seines. *Commercially:* accidentally in trawls. Often found on beaches, probably after being discarded from a fishing boat. This fish is almost certainly poisonous.

Further Reading

Ayling, T., Cox, G.J. 1982; *Collins Guide to the Sea Fishes of New Zealand.* Collins, Auckland. (Revised edition 1984)

Doak, W. 1972: *Fishes of the New Zealand Region.* Hodder and Stoughton, Auckland. (Revised edition 1978)

Doak, W. 1991: *Wade Doak's World of New Zealand Fishes.* Hodder and Stoughton, Auckland.

Doogue, R.B., Moreland, J.M., Heath, E.W. 1960: *New Zealand Sea Angler's Guide.* Reed, Wellington. (Revised editions 1961, 1982)

Francis, M. 1997: *Coastal Fishes of New Zealand. An Identification Guide.* Reed, Auckland.

Graham, D.H. 1953: *A Treasury of New Zealand Fishes.* Reed, Wellington. (Revised edition 1956, reprinted 1974)

Paul, L.J. 1986: *New Zealand Fishes.* Reed Methuen, Auckland.

Paul L.J. & Moreland, J. 1993: *Handbook of New Zealand Marine Fishes.* Reed Books, Auckland.

Paulin, C., Stewart, A., Roberts, C., McMillan, P. 1989: *New Zealand Fish. A Complete Guide.* National Museum of New Zealand Miscellaneous Series No. 19. GP Books, Wellington.

Paulin, C., Roberts, C. 1992: *The Rockpool Fishes of New Zealand. Te ika aaria o Aotearoa.* Museum of New Zealand Te Papa Tongarewa, Wellington.

Index of Common Names

Index of Scientific Names